Computerised
Environmental
Modelling

Computerised Environmental Modelling

A Practical Introduction Using Excel

J. HARDISTY,
D. M. TAYLOR
and
S. E. METCALFE
The University of Hull, UK

JOHN WILEY & SONS
Chichester · New York · Brisbane · Toronto · Singapore

Copyright © 1993 by John Wiley & Sons Ltd,
Baffins Lane, Chichester,
West Sussex PO19 1UD, England

Reprinted August 1994

Other Wiley Editorial Offices

John Wiley & Sons, Inc., 605 Third Avenue,
New York, NY 10158-0012, USA

Jacaranda Wiley Ltd, G.P.O. Box 859, Brisbane,
Queensland 4001, Australia

John Wiley & Sons (Canada) Ltd, 22 Worcester Road,
Rexdale, Ontario M9W 1L1, Canada

John Wiley & Sons (SEA) Pte Ltd, 37 Jalan Pemimpin #05-04,
Block B, Union Industrial Building, Singapore 2057

Library of Congress Cataloging-in-Publication Data

Hardisty, J. (Jack), 1955–
    Computerised environmental modelling : a practical introduction
using Excel / J. Hardisty, D.M. Taylor, and S.E. Metcalfe.
        p.   cm.
    Includes bibliographical references and index.
    ISBN 0-471-93822-X (pbk.)
    1. Environmental sciences—Data processing.   I. Taylor, D. M.
(David M.)   II. Metcalfe, S. E. (Sarah E.)   III. Title.
    GE45.D37H37   1933
    628.5′01′13–dc20   93-18848
                                                                CIP

British Library Cataloguing in Publication Data

A catalogue record for this book is available from the British Library

ISBN 0-471-93822-X

Typeset in 11/13pt Palatino from authors' disks by Text Processing Department,
John Wiley & Sons Ltd, Chichester
Printed and bound in Great Britain by Biddles Ltd, Guildford, Surrey

Contents

Part III : Examples of Environmental Models

Part IV : Appendices

Preface

The geographical, earth and environmental sciences are currently undergoing two phases of development which, for undergraduates, are generating two related but separate problems.

On the one hand, the management of the environment is developing into a rigorous and demanding analytical science and there is, therefore, a growing expectation that departments will produce graduates trained to deal with environmental issues. The complexity of the environment often necessitates the use of sophisticated environmental models and the use of such models, particularly computer models has become a part of the undergraduate learning process. The second problem arises from these developments and is concerned with 'how should the environment be taught with a vocational emphasis?' and 'how should modelling be taught at all?'.

We have tried to provide one coherent set of answers to these problems in this book. The environment is taught through the use of numerical models (for these will be the analytical tools of management in the 1990s) and numerical modelling is taught through the use of microcomputer-based spreadsheets (for these will continue to be the computer tools of management in the 1990s). This answer is more subtle than it seems. The use of an industry standard spreadsheet (we have worked with Microsoft Excel, but others such as Works or Lotus are equally useful) with all of its built-in functions and graphical capabilities obviates the need for language-specific computer models and the students do not need to program before they can model. The teaching of a computer language as a prerequisite to computer modelling has severely limited such courses and such textbooks in the past.

Computerised Environmental Modelling, which is based upon a current undergraduate course in the School of Geography and Earth Resources at the University of Hull, attempts to address these problems by presenting a readable (we hope) and tested (we know) introduction to the subject through an easy to use software system. It should teach students how to design and construct quite sophisticated environmental models on desk-top microcomputers.

Finally, although we accept the strictures of collective responsibility, it may be useful to report that each of us was principally responsible for different chapters. JH worked up Part II and Chapters 11 and 12; DMT wrote Chapters 1, 13 and 14 and SEM wrote Chapters 2, 15 and 16.

Jack Hardisty
David Taylor
Sarah Metcalfe

Part I
ENVIRONMENTAL MODELLING

1
Environmental Systems

*I do not know how the parts are interconnected, and how each part accords
with the whole; for to know this it would be necessary to know the whole
of nature and all of its parts*

Baruch Spinoza 1632–1677

INTRODUCTION

A system can be viewed as a box within which there is a set
of interrelated components. The box has inputs and outputs, the
levels of which depend upon the permeability of the box's walls
(boundaries), or the degree of 'openness' of the system. In the
environmental sciences, systems can be visualised over a range of
scales. For example a tree leaf, together with all the microscopic
and macroscopic fauna and flora (i.e. the biota) that live on and
in it is a system (as the main focus is living matter, this type of
system is commonly classed as an 'ecosystem'). In the case of the
leaf ecosystem, the system's boundaries are the outermost surface
of the leaf and its biota. Inputs to the system are then sunlight,
water and nutrients translocated through the plant to the leaf;
outputs include reflected energy and detrital material falling to the
woodland floor. The system's components (leaf surface, fauna and
flora) are interrelated through the links of the food web; for example
herbivorous insects feed off the leaf before, through being consumed
themselves, passing on part of the energy they obtain to carnivorous
insects. At the opposite end of the scale the Earth's climate can be

viewed as a system (although it is more often viewed as a series of smaller, nested subsystems). In this the system's boundaries are the Earth's surface and space, the inputs and outputs are energy, the components include the main water bodies and the atmosphere and the inter-connections are energy transfers.

Further descriptions of environmental systems are not difficult to find in the literature. Conveniently, in view of the context of this chapter, Johnston (1989) uses the Isle Royale island ecosystem as an example of a much simplified environmental system. The Isle Royale example, originally put forward by Rykiel and Kuenzel (1971), is displayed as a set of boxes (components) and arrows (interrelationships) as shown in Figure 1.1. The main basis for the system comprises three interrelated system components: plant biomass, moose population and wolf population. These three are closely linked; hence, by increasing the plant biomass the moose population and wolf population are increased directly and indirectly, respectively. The three are also affected by other factors, which are lumped together as two additional system components: physical (atmospheric and soil) and biological (competition and predation from other herbivores and carnivores) processes.

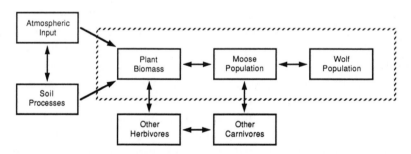

Figure 1.1   *A simplified ecosystem: Isle Royale (after Johnson, 1989; adapted from Rykiel & Kuenzel, 1971)*

The process of breaking down our otherwise highly complex environment into a number of more or less discrete systems, each of which is comprised of inputs, outputs and interconnected components, in order that it might more easily be studied is known as the 'systems approach'. The systems approach thus allows the student to focus on only that which is of direct interest (in many cases this is the nature and level of the interrelationships between components); everything else is ignored. In other words, the systems approach maintains that, in order to understand the functioning of the environment, it is not necessary to know the *whole* of nature and *all* of its parts.

Two assumptions underpin the systems approach. First, that it is actually possible to subdivide the real world into discrete, contained or partially contained, functioning systems, and second that it is possible to determine the various inputs and outputs and interrelationships between system components. In the Isle Royale example these assumptions may hold true; being a clearly defined island it should be relatively easy to define and isolate the ecosystem's boundaries, inputs, outputs and components. However, in the real world this is not always the case; virtually all ecosystems grade into (through an ecotone) and interact with adjacent ecosystems. Hence the real world is much more complex than our examples would immediately suggest, with system components and system inputs and outputs, even if they are all definable, usually being shared with other systems. Similar problems arise when the systems approach is adopted in other branches of the biological and geographical sciences. In fact truly isolated environmental systems are the exception rather than the rule and are not frequently encountered outside the laboratory.

Gregory (1985), in his summary of the development of the systems approach in the various branches of science, claims that it was well established in physical geography by 1970 and goes on to describe its progressive adoption by biogeographers, pedologists, climatologists and geomorphologists. Part of the attraction of the systems approach to environmentalists and geographers is that, through its provision of a common methodology and nomenclature, greater inter- and intra-disciplinary study is facilitated. As Cooke (1971) opines, 'In terms of integrating physical geography with related disciplines, a systems approach undoubtedly succeeds.' During the early period of its adoption in physical geography the systems approach was focused very much on investigations of environmental systems perceived as being in a state of static equilibrium. In these the dominant system interrelationships and controls were assumed to be, respectively, linear relationships and negative feedback (homeostatic) mechanisms (discussed in Chapter 9). Thus changes in system components were assumed to be continuously and proportionally related to changes in their controlling variables, whilst small departures from the equilibrium state (or the 'norm') were reversed by negative feedback. Very little emphasis was placed on attempting to understand how systems evolve, or develop, through time and space. Thornes (1987) calls this the 'dynamical behaviour' of systems. Indeed Thornes reflects that the earlier assumptions behind the systems approach have limited its applications and success in physical geography. He goes on to point out that, '...it seems ironic that the richer and yet more profitable side of the adoption of systems thinking, the understanding of dynamical behaviour, has been almost entirely neglected, even in university teaching and research.'

Thornes (1987) describes the developments in systems analysis and theory that have led to changes in emphasis in environmental system design during the last two to three decades. To illustrate this he cites the changes that have been made in the study of alluvial river channel systems in order to accommodate new ideas and discoveries. For example it is now appreciated that rather than a single, stable equilibrium state, environmental systems generally have several equilibrium states. In the case of alluvial river channel systems, the system may in some cases require only a relatively small change in one of the controlling variables (such as slope or discharge) to cause a sudden flip from one apparently stable state (e.g. a meandering channel) to another (e.g. a braided channel). Examples such as this, where the state of natural systems appears prone to abrupt and dramatic change, highlight a further problem with the static equilibrium approach in that in some cases relationships between system components are not dominated by negative feedbacks, but by positive feedback processes. We can well illustrate this through the alluvial river channel example: a relatively small change in, for example, vegetation cover in the catchment might result in a dramatic, even violent, irreversible response, such as a landslip. This implies nonequilibrium rather than stable equilibrium behaviour. Finally, Thornes criticises the static equilibrium approach because it focuses on systems in their present state, with little regard being paid either to their origin (e.g. what processes bring about the initiation of alluvial river channels?) or to their evolution (i.e. change, be it gradual or sudden) through time.

Since the early 1970s there has been a shift in emphasis in the environmental sciences towards the *dynamical systems approach*. Important aspects of this are that it attempts to incorporate the history of the system under study and that it focuses on change and the causes of instability. As the majority of our environmental problems at present seem to be dominated by catastrophic responses and positive feedback mechanisms, a dynamical systems approach is a useful tool in environmental management and monitoring. This will be returned to later in the present chapter and in Part III.

ENVIRONMENTAL SYSTEMS ANALYSIS

The study of the composition and functioning of systems is known as 'systems analysis'. Systems analysis is of relevance in the present book because it is closely tied to modelling. Indeed systems analysis and computer modelling are virtually synonymous in the environmental sciences; system construction is often an early stage in computer modelling of the environment, and systems analysis requires the development of a model, either verbal, mathematical or physical, in order to simplify the original complexity present in the environment. According to Huggett (1980) there are 4 distinct phases or stages in systems analysis:

1   *The lexical phase.*   This involves three steps: the actual defining of the system's boundaries; the choosing of the system's components (hereafter referred to as 'state variables'); and the estimation of the value (state) of the state variables. Huggett calls these steps system *closure*, *entitation* and *quantitation*, respectively (the last two terms are from Gerard (1969)). It is worth noting that, once defined, the state variables are subject to change, both temporally and spatially. Indeed it is changes in state variables that characterise Thornes' dynamical systems. Furthermore, determining the value of the state variables at set intervals of time provides a means of monitoring

dynamical systems, a subject which will be returned to later on in this chapter. Thornes defines four ways in which state variables can change through time: *damped* (where state variables stay at, or fluctuate around, a fixed level); *explosive* (where state variables move progressively away from an initial value); *periodic* (where state variables move between extremes through time); and *unsystematic* (where there is no perceptible pattern of change through time, i.e. change is chaotic).

2 The parsing phase.   This involves defining the relationships between the state variables of the system, either verbally, mathematically or physically. The models described in this book all involve mathematical relationships between state variables. The relationships are either *deterministic* or *stochastic* (the latter include a random or probability component). Deterministic mathematical relationships are usually in the form of either *differential* or *difference* equations; the former refer to continuous relationships, e.g. the increase in a population's size with time, and the latter to discontinuous, or step, relationships, e.g. total population size after one year.

3 The modelling phase.   This involves two steps; the first is the actual model construction, which requires the elucidation of how changes in controlling variables affect the state variables. The second is the running (i.e. *operationalising*) of the model.

4 The analysis phase.   In this the model is *validated* (see Chapter 10). This involves comparing the results of the model with the observed data in the field. In many cases there is only a poor correlation between the two; if so, then phases 3 and 4 must be repeated, once the model has been re-designed.

There are several ways of depicting systems on paper. Perhaps the most common method in physical geography is as a flow chart or network diagram, using the symbols introduced by Chorley and Kennedy (1971) (Figure 1.2).

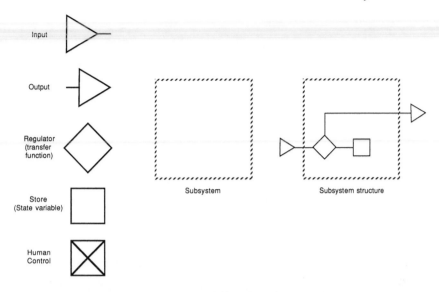

Figure 1.2   *Symbols commonly used in systems design (after Chorley & Kennedy, 1971)*

THE CLASSIFICATION OF ENVIRONMENTAL SYSTEMS

As a result of the interrelated nature of environmental systems and because of the range of disciplines (and hence scientists) that have adopted the systems approach, a totally acceptable, all-embracing classification of system types is not available. Indeed Huggett (1980) states that the '...classification of system types is one of the most confusing parts of the systems literature'.

Two means of classifying environmental systems are according to their geographic location and their degree of 'openness'. For example, systems can be classified with reference to their position in the atmosphere, biosphere, hydrosphere or lithosphere. Alternatively they may be classed as open, semi-open or closed, according to the level of system inputs and outputs. None of these classifications has, however, gained universal acceptance.

Chorley and Kennedy introduced a more widely accepted classification of environmental systems in 1971, based upon the degree of internal system complexity (Figure 1.3). Accordingly, *morphological systems* (Figure 1.3(A)) are the least complex type. In

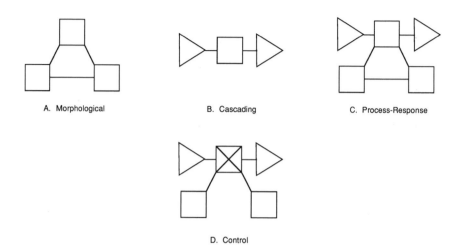

Figure 1.3 *Classification of systems (after Chorley & Kennedy, 1971)*

these, the emphasis is placed upon the relationships between state variables, and upon the relative strengths of these interrelationships. Such systems are of obvious benefit when determining the possible effect of changes in one system state variable on the others. For example, it is possible to predict the impact on the moose population of increasing the number of predatory wolves on Isle Royale.

At the next level of increasing system complexity are *cascading systems* (Figure 1.3(B)). In these the system components are linked by flows of mass or energy or a combination of the two. An example is the flow of energy through an ecosystem; solar energy entering the system is either temporarily stored as carbohydrate at various trophic levels, or converted and released as long-wave energy. More complex still are *process-response systems* (Figure 1.3(C)), as these comprise parts of cascading and morphological systems. Johnston (1989) proposes the relationship between wave power (an energy input and hence part of a cascading system) and the angle of slope of a beach (a structural output and hence part of a morphological system) as an example of a process-response system. Process-response models are described in Chapter 7.

The fourth type of environmental system described by Chorley and Kennedy are *control systems* (Figure 1.3(D)). These are similar to the process-response systems, but include a human controlling factor. Thus to convert our beach process-response system into a control system, we need only add a human controlling factor, such as a coastal defence component. As a result of this strong emphasis on human influence, control systems have useful applications in environmental management and the solving of environmental problems. It is these applications, together with the monitoring of dynamical systems, which form the main focus of the rest of this chapter.

ENVIRONMENTAL MONITORING, MODELLING AND MANAGEMENT—THE SYSTEMS APPROACH

The objective of environmental management is to optimise the environment, so that it better meets the requirements of an ever-increasing human population, or to reduce the environmental impact of human activity. The first may be achieved, for example, by productive agricultural and forestry systems and implies some level of deliberate human modification of the environment. The second is attained through, for example, pollution control; here the emphasis is often on lessening the impact of a human-induced, positive feedback process.

The systems approach, when combined with computer-based models, is a powerful tool in the management of natural resources such as crops (e.g. Lakshminarayan *et al.*, 1991), forests (e.g. Usher, 1976) and rangelands (e.g. Starfield & Bleloch 1986). Davis (1992) presents a model for the sustainable utilisation of wildflower communities on the Fynbos in South Africa, a natural resource that has received little attention in the past from systems analysts. The wildflowers have both commercial and biological values; each year more than 1×10^6 kg of fresh plant material is collected and exported, mainly to markets in western Europe, whilst 68% of the approximately 8600 plant species are endemic to the area. Davis' model (named 'VELDFLOW') is designed to predict the effect of increased levels of harvesting on future wildflower populations. According to Davis, the model should assist the 'decision-making processes of an entrepreneur wildflower producer whose economic gains are measured against the impact of harvesting on utilized plant populations'. Perhaps not surprisingly, economic returns can be raised by increasing the amount of wildflowers harvested up to a certain level without affecting future supplies. Beyond this level, or *threshold*, further harvesting, through its deleterious impact on pollination, seed bank stores, soil texture, etc., reduces the following year's harvest, and thus the following year's economic return.

A systems approach has applications in the monitoring of environmental systems. For example, Wiersma, Otis and White (1991) describe how simple, predictive models can be used to assist in the monitoring of levels of pollutants (in this case Martha's Basin, Glacier National Park). Their methodology closely follows Huggett's (1980) four-stage procedure for systems analysis, described earlier in the chapter, except that the monitoring objectives are clearly stated at the outset. One of the problems with environmental monitoring *per se* is that there is a tendency for the results merely to illustrate what has happened in the past. By adopting a systems approach it is also possible to describe what the future may hold in store; for example, predictions on the effect of continuing high pollution levels on the components of a lake system can be made according to a 'business as usual' scenario.

An example of the systems approach assisting in the solution of environmental problems comes from the work of Brian Moss. Moss (1987) determined the interrelationships between, and impacts of, changes in land use, water quality, visitor pressure and habitat diversity in the Norfolk Broads. One beneficial consequence of his research was that, in highlighting the effect of phosphate-rich sewage effluent on biological communities and channel morphology, it forced the Anglian Water Authority to improve the treatment of effluent entering the Broads system (Owens & Owens, 1991).

There are many examples in the literature in which environmental managers have utilised a systems approach, in addition to those described above. It is highly probable that this use will continue, especially as human expectations and population levels show no sign of even stabilising, let alone declining, and demands upon the environment, and hence environmental managers, continue to increase. However it must be remembered that the usefulness of the approach is limited by the great complexity and innumerable inter-connections that prevail within the environment. As Bennett and Chorley (1978) suggest in the concluding chapter of *Environmental Systems*, despite all the gains and developments made since it first became widely accepted, the systems approach is still not 'a universal panacea for all our philosophical and technical problems'. Thus it is not yet able to provide the absolute answer to Spinoza's dilemma, although the systems approach can provide a useful basis from which to study and understand the functioning of the environment.

END NOTE

In many instances the application of the systems approach, in which computer models form an increasingly important part, has proved useful to the study and understanding of our environment. The rest of the book now concentrates on computer-based environmental models: Chapter 2 discusses environmental modelling *per se*; Part II provides a methodology through which environmental models can be constructed and validated; and Part III gives some examples of environmental models using Excel.

REFERENCES

Bennett, R.J. & Chorley, R.J., 1978. *Environmental Systems Philosophy, Analysis and Control*. Methuen, London.

Chorley, R.J. & Kennedy, B.A. 1971. *Physical Geography. A Systems Approach*. Prentice-Hall International, London, 370 pp.

Cooke, R.U., 1971. Systems and physical geography. *Area*, **3**, 212–216.

Davis, G.W., 1992. Commercial exploitation of natural vegetation: an exploratory model for management of the wildflower industry in the Fynbos biome of the Cape, South Africa. *Journal of Environmental Management*, **35**, 13–29.

Gerard, R.W., 1969. Hierarchy, entitation and levels. In: L.L. Whyte, A.G. Wilson & D. Wilson (eds) *Hierarchical Structures*. Elsevier, New York, pp. 215–230.

Gregory, K.J., 1985. *The Nature of Physical Geography*. Edward Arnold, London.

Huggett, R., 1980. *Systems Analysis in Geography*. Clarendon Press, Oxford.

Johnston, R.J., 1989. *Environmental Problems: Nature Economy and State.* Belhaven Press, London.

Lakshminarayan, P.G., Atwood, J.D., Johnson, S.R. & Sposito, V.A., 1991. Compromise solution for economic–environmental decisions in agriculture. *Journal of Environmental Management*, **33**, 51–64.

Moss, B., 1987. The Broads. *Biologist*, **34**, 7–13.

Owens, S. & Owens, P.L., 1991. *Environment, Resources and Conservation.* Cambridge University Press, Cambridge.

Rykiel, E.J. & Kuenzel, N.T., 1971. Analog computer models of The wolves of Isle Royale. In: B.C. Patten (ed.) *Systems Analysis and Simulation in Ecology*, vol. 1. Academic Press, London, pp. 513–541.

Starfield, A.M. & Bleloch, A.L., 1986. *Building Models for Conservation and Wildlife Management.* Macmillan, New York.

Thornes, J.B., 1987. Environmental Systems. In: M.J. Clark, K.J. Gregory & A.M. Gurnell (eds) *Horizons in Physical Geography.* Macmillan, London, pp. 27–46.

Usher, M.B., 1976. Extensions to models, used in renewable resource management, which incorporate arbitrary structure. *Journal of Environmental Management*, **4**, 123–140.

Wiersma, G.B., Otis, M.D. & White, G., 1991. Application of simple models to the design of environmental monitoring systems: a remote test site. *Journal of Environmental Management*, **32**, 81–92.

2
Introduction to Modelling

And they'll build systems dark and deep,
And systems broad and high;
But two of three will never agree
About the reason why

The Wise Men of Gotham
Pindar

INTRODUCTION

It seems that every book about modelling, or concerning the application of models, includes an introductory chapter with at least one definition of what a model is and describing a range of types of models (e.g. Gilchrist, 1984; Kirkby *et al.*, 1987). Rarely do the authors produce the same definition or the same classification of model types. The reader may, therefore, wonder why this practice is being continued, why not simply provide a list of references and leave the reader to their own devices? Apart from the rather flippant answer of 'habit', the response must be to provide the background and context for the present text. All the authors of this book work in a geography department, although not all trained as geographers. As a result, our view of modelling is perhaps best seen in the context of attitudes to, and the development of, modelling in that discipline. The case for a model-based paradigm (academic framework) in geography was first mooted by Haggett and Chorley in 1965 at a time when the subject was striving to become more quantitative. This approach was stated explicitly in the later publication *Models in Geography* (Haggett & Chorley, 1967). Developments of modelling, examples of applications and criticisms of the whole idea over the period since the Madingley meeting have been brought together by Macmillan (1989). Whilst rooted in geography, the latter volume emphasises the need for environmental modelling in its widest sense. Physical, hydrological and biological systems have all been 'modelled' in some way or another, with the recognition of an increasing need to include the interaction of 'physical' and 'human' systems.

The idea of computer modelling can be a frightening one in its apparent difficulty and complexity. In this book we set out to show that Computerised Environmental Modelling can be explored from a fairly basic starting point, without the need to master a traditional programming language. To return to tradition, however, this chapter will attempt to provide a review of the rationale for and means of modelling. Some of the dangers of the use of mathematical models will also be pointed out.

WHY MODEL?

Generally, models represent a simplification of reality, and hence a means of 'getting to grips' with systems whose spatial scale or complexity might otherwise put them beyond our physical or mental grasp. This can be particularly valuable for those interested in the workings of environmental systems which tend to be inherently complex. In spite of the simplification, it is to be hoped that models retain the significant features or relationships of reality. Hence, all models are subjective, as the modeller chooses those elements of the real world that should be included in the model, as well as the mode of representation. In some ways the process of modelling may be likened to the act of drawing a plant or animal; it makes you look very closely at the visible structure and may also increase recognition of the significance of that which is unseen. Modelling itself may, therefore, provide a stimulus for thought. Models are used to describe, explore and analyse how a system works. Increasingly too modelling has become part of the process of planning and policy with the need for good communications between those who develop the models and those who use them. This takes us into the realms of using models for predictive purposes, to 'what if?' scenarios. In fact Haines-Young and Petch (1986) have got to the stage of defining what a model is in terms of an ability to generate predictions. Whatever the definition, modelling is not, and should not be, a substitute for thinking and will only be effective if combined with an interest in and knowledge of the system being modelled. As Gilchrist (1984) has indicated 'The modeller begins with ideas, experience and a desk full of relevant literature and data.'

MODELS—FROM PLASTICINE TO PASCAL

'Surprisingly, after more than two decades of modelling, we are still debating what a model is' (Macmillan, 1989). In the light of this statment, it may seem foolish to attempt any sort of definition of what models are. Each author produces his or her own definition (see above). In very general terms, however, we might revert to that proposed by David Harvey in an earlier incarnation 'a model is a temporary device to represent what we think a structure may, or ought, to be' (Harvey, 1969). Most models are approximations of reality, with a greater or lesser degree of simplification; if not then one might pose the question 'why model?' Because of this basic principle, modellers are necessarily highly selective of the information they actually use. The question then arises of the basis on which features/relationships are selected: intuition, experimentation or theory? The relationship between models and theory is one which has attracted a lot of attention. In his discussion of this, Harvey (1969) indicates that in general terms the most productive links between a theory and a model should be mediated through the imagination:

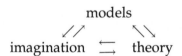

Broadly speaking, models may be classified on the basis of their chronological position in relation to theory development as *a priori* or *a posteriori* models. For *a priori* models, an idea or construct results in the development of a model which may, in turn, lead to the formulation of a theory through the process of hypothesis testing. Under such circumstances, systems may be investigated in the absence of an existing, complete theory and one may subsequently be developed. Harvey points out that this is the most common situation in relation to geographical (and environmental) systems. The alternative situation is one where observation has already resulted in the development of a theory, which may then be explored using an *a posteriori* model. As knowledge progresses, a model may change from being *a priori* to *a posteriori* in type. Over the same period, the function of the model may alter from that of 'picturing', to ordering and, finally, to explaining.

The classification of models into 'types' is almost as confused as the problem of definition. Haggett and Chorley (1967) review some of the bases on which models could be classified. A basic division can be made between descriptive and normative models. Descriptive models offer a stylised portrayal of reality, with either an emphasis on equilibrium structural features (static), or on changes in processes and functions through time (dynamic). Normative models involve the use of analogues, applying a better known situation to a lesser known one, and are usually used in a predictive sense. Other approaches are to consider the types of materials used or the nature of representation: conceptual, theoretical or symbolic. In symbolic models reality is represented by a logic system, for example mathematical equations. Mathematical models, therefore, fall into this category.

In very general terms then, three types of model can be recognised:

1. *Natural analogues.* The use of actual events or objects occurring in different times or different places, to help to explain what has, is or will happen to a particular system. Examples of this approach include:
   (a) The principle of uniformitarianism (established, if not named, by Lyell, 1830–1833) of 'the present is the key to the past' in the application of current processes to explain landform evolution.
   (b) The use of old systems (e.g. landforms) to predict the future state of present landforms. Denudation chronology is an example of this form of analogue.

(c) Using events in one place as a model for what is happening elsewhere.

2. *Hardware or physical models*. In this case, a range of materials (often natural) are used. Three types may again be identified:
   (a) Models which use the materials of the natural system and are geometrically and dynamically similar. Sometimes a portion of the natural system (e.g. beach) is itself used.
   (b) Natural materials are again used; dynamically similar but geometrically disimilar to the natural system.
   (c) Simulation of the dynamic behaviour of the system using different materials.
3. *Mathematical models*. A range of approaches including deterministic, stochastic (or probabilistic) and optimisation models. These are discussed in more detail below.

Chorley (1967) produced a model classification which included a number of the elements described above. His classification is reproduced in Figure 2.1. Although identifying three main categories of model (analogue, physical, general system), this classification

1. Natural Analogue System
   (a) Historical analogue
   (b) Spatial analogue

2. Physical System
   (a) Hardware model
       (i) Scale
       (ii) Analogue
   (b) Mathematical
       (i) Deterministic
       (ii) Stochastic
   (c) Experimental design

3. General System
   (a) Synthetic
   (b) Partial
   (c) Black box

Figure 2.1 *Chorley's (1967) classification of model types*

was developed from the idea that all models were analogues of some kind, although differing in style (Chorley, 1964). Models in category 1 (natural analogue system) have been described above. An example of this approach has been the idea of using previous warm periods in the geological record as analogues for the effects of enhanced greenhouse warming in the future (Budyko, 1987). The limitations of this have been described elsewhere (IPCC, 1990). The second category (physical system models) includes a range of more usual scientific strategies and elements of it will be explored in due course. The idea of looking at models in terms of systems and causal mechanisms is reflected by category 3 (general systems). In this case, the emphasis is on replicating the functioning of the system by reproducing it, rather than concentrating on analysing the system (categories 1 and 2). In synthetic (or artificial) system models, reality is simulated in a structural way with internal processes specified. The name 'white-box' model is applied to models of this type. In practice the full range of processes operating in a system are hardly ever known. In partial system models, the emphasis is on workable relationships, where it is possible to derive results without complete knowledge of the internal workings of the system being studied. Models of this type are also known as 'grey box' models. In the final category, 'black-box' models, inputs result in outputs but with no knowledge of what actually goes on in between. Relationships may be expressed mathematically, but processes are not (cannot) be expressed explicitly. Refer to Chorley (1964) for further details!

MATHEMATICAL MODELS

The models presented in this book all fall into this category, although it includes a wide range of modelling approaches. The mathematical representation of the dynamics of a real life system is called simulation modelling. The application of simulation models to environmental problems has been reviewed by Frenkiel and Goodall (1978).

The major types of mathematical models will now be listed; their relationships are illustrated in Figure 2.2. In deterministic mathematical models for a given input, there is only one output,

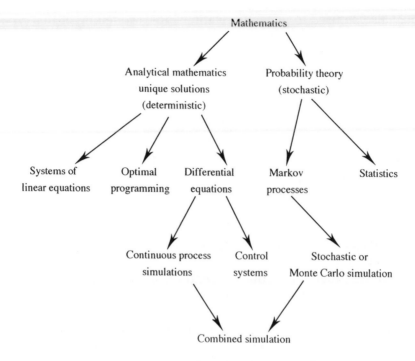

Figure 2.2

i.e. unique solutions are obtained. The simplest type of relationship between two variables is a linear one, where the relationship between x and y is proportional. A simple process-response model applies (see Chapter 7). A basic least-squares regression equation ($y = ax + b$) is an example of such a linear relationship. Linear systems are easy to deal with because linear equations are soluble: there is an answer.

In stochastic mathematical models, by contrast, there are a range of possible outcomes for any one input reflecting randomness, or uncertainty, in the system. There is no single answer. In models, this uncertainty is included through the use of probability. One approach is to use the Markov chain, where the probability of an event occurring is dependent only on the event preceding it. Another strategy is to use the Monte Carlo method where a solution for a problem is developed by carrying out sampling experiments on a stochastic model. The outcomes are, therefore, independent of previous states of the system. The direct calculation of probabilities may also be used to generate relationships. Modelling of this type is explored further in Chapter 8.

If the relationship between variables is complex (*i.e.* non-proportional), then a system is described as non-linear. Such systems are probably the norm in the natural environment, where a range of feedbacks (positive or negative) can be seen to operate. The formulation of models of non-linear systems involves both deterministic and stochastic elements.

It may be a requirement of the modelling exercise to maximise, or minimise, some aspect of the output of the system to meet a set of criteria. Models of this type are called optimisation models.

The use of mathematics to express relationships implies the carrying out of calculations. Whilst the determined modeller may be prepared to hammer away at a calculator, the usual prefered strategy today is to write down all the calculation instructions for a computer; that is to write a computer program. Later in this book we present a number of sets of calculations (programs) to represent the functioning of a number of different natural systems. In the real world, processes operate as continuous functions; in modelling, however, these functions are reduced to discrete numbers (see Chapter 11). Although it is possible to describe processes changing smoothly over time using differential equations, the approach adopted here is the simpler one of applying difference equations which represent processes 'jumping' from state to state. Whilst this eases calculation, it must be remembered that it may represent a significant departure from the operation of the real system.

THE MODELLING PROCESS

The first step in any modelling exercise is to decide whether a model is necessary and, if so, to decide what type is the most appropriate. Some of the options have been described in the previous sections. The choice of model type will depend on the aims of the exercise, the nature of the system to be modelled and the level of understanding of the processes operating in the system. The availability, or unavailability, of appropriate data may be a major constraint. Ideally, data should be collected in the framework of hypothesis testing within which modelling can play a vital part. The model may provide the basis for testing, so that data collected under such circumstances would be appropriate for the model. In modelling environmental systems the most appropriate type of model is usually a mathematical (*sensu lato*) simulation model.

The next steps are then to identify an appropriate model structure, to estimate the parameters which will characterise the model and finally to validate the model. Parameter estimation and model validation pose particular problems with respect to modelling environmental systems. The complexity of such systems may make it difficult to identify the relationships and mechanisms upon which they depend. Parameterisation is a way of expressing an assumed relationship between variables in mathematical form. Estimates of some parameters may be based on observations or laboratory experiments (empirically derived) or based on scientific principles (theoretically derived). At worst, a process or group of processes may simply be left out (null parameterisation). The level of understanding of the system to be modelled and the uncertainties involved therefore provide the essential basis for model development. Models can only be as good as the knowledge or understanding at the time of their construction. As Thornes (1989) has put it, the credibility of models hinges upon the credibility of their science base. Once a model is running, the values of parameters may be adjusted ('tuned') during calibration to make the model output match observations more closely. In very complex systems it may simply not be practical to spend a lot of time setting and resetting parameters. One solution to this is to carry out a sensitivity analysis. This involves deliberately changing parameter values to assess their effects on model outputs. A relatively small number of parameters may prove to be those with most influence and as a result these should receive the greatest attention. Adjustment of parameter values should not take them beyond the bounds of reality just to improve the match between output and observations. If this occurs, something significant is clearly missing from the model.

The ultimate test of a model is validation, in other words comparison of the model with the real world system it is supposed to portray. In principle, validation requires the use of a completely different set of data from that used to construct the model in the first place. Irrespective of how carefully validation is carried out, there will always be some degree of uncertainty in model output. Models are only really valid within the bounds of their original context, for example if a model for a catchment is developed under given climatic and land-use conditions, changes in either of these (e.g. afforestation) will undermine the model's validity.

Modelling, like most things, has its limitations and pitfalls. The desire to simplify, to make modelling possible, may be taken too far resulting in 'throwing the baby out with the bathwater' (Haggett & Chorley, 1967). Models must also remain realistic in the way that systems are represented. If due care is not taken then the predictions produced by the model will be invalid (a problem called 'suggestiveness'), although perhaps retaining a spurious air of respectibility. This is a particular problem now that many large computer models are too complex for all but the specialist to understand and criticise.

CHOOSING A PLATFORM AND A LANGUAGE

The development of modelling as a tool for explanation and prediction has closely followed developments in computing—both hardware and software. Early computer programs were as simple as the computers were large and cumbersome. When the earliest computers were developed few people recognised their potential as anything other than giant calculators. One who did was John von Neumann of Princeton who saw that modelling weather might be an ideal task for them. In 1960 Edward Lorenz built a simple climate model with three variables (temperature, pressure and wind speed) which were linked by 12 parameters in a deterministic way (Gleick, 1988). Over time, climate models have become more and more complex running on bigger, faster computers (Henderson-Sellers & McGuffie, 1987). Modern General Circulation Models (GCMs) now model the atmosphere over a global grid (about 100 km^2), through several vertical levels, with up to a million variables and 500 000parameters. The variables are integrated through time using three-dimensional equations of motion and thermodynamics. The most recent development has been that of fully coupled ocean-atmosphere models, which stretch even the latest computers. Failures of models to provide adequate predictions can now be blamed on inadequate computing power. A fine example of this was the failure of the Meteorological Office to predict Britain's worst storm in 300 years when it occurred on the 15–16 October 1987. This failure was blamed on an 'outdated', single processor Cyber 205 computer.

On the basis that bigger may be better, the idea has been put forward that numerical calculations could provide full solutions to all the complexities of real life situations. Is there, therefore, still a role for highly simplified models? The answer appears to be 'yes', as they can provide a means of understanding and assimilating great complexity, possibly revealing a system's underlying structure. They also have the great advantage that the assumptions and limitations of a simple model can easily be understood by a range of people, not only the person who wrote the model. As computer models become more widely used in universities, businesses and government departments, this aspect is very important. There is still a case for starting with the simplest model first (Gilchrist, 1984).

Although mainframe computers have become more powerful, the greatest development has been in the field of desk-top computing. IBM PCs (and their clones) and Macintosh computers can now pack as much computing power as a small mainframe, but with the advantage of being smaller and usually much more user friendly (Radford, 1991). As a result, more and more people do their computing in this environment. Following this shift in 'platform' (from mainframe to PC) has been a move away from the need to master conventional programming languages (e.g. FORTRAN, PASCAL, COBOL, and even BASIC) within less than cooperative operating systems. Although a wide range of models for Macintoshes and PCs are still written in these languages, it is also possible to 'program' in a spreadsheet environment. This is the approach that we have adopted in this book using the Excel spreadsheet. We hope that modelling in Excel will introduce students and other users in a relatively painless way to the stimulating world of computerised environmental modelling.

FINAL THOUGHTS

Computer modelling now offers an accessible and rapid means of exploring the operations of natural (and human) systems, with ever-increasing emphasis on the use of models in prediction, policy and planning. It is, perhaps, easy to get carried away by the technology, to forget all the assumptions and limitations built in to every computer model, no matter how complex. Even the mighty (Jim Hansen, Director of NASA's Goddard Institute of Space Studies) can get their knuckles rapped if they do appear to forget (Kerr, 1989). Finally, and in common with all other writings about modelling, remember the old adage 'Garbage in, garbage out'.

REFERENCES

Budyko, M., Ronov, A.B. & Yanshin, A.L., 1987. *History of the Earth's Atmosphere*. Springer-Verlag, Berlin (English translation).

Chorley, R.J., 1964. Geography and analogue theory. *Annals of the Association of American Geographers*, **54**, 127–137.

Chorley, R.J., 1967. Models in Geomorphology. In: R.J. Chorley & P. Haggett (eds) *Models in Geography*. Methuen, London, pp. 59–96.

Frenkiel, F.N. & Goodall, D.W., 1978. *Simulation Modelling of Environmental Problems*, SCOPE 9. John Wiley & Sons, Chichester.

Gilchrist, W., 1984. *Statistical Modelling*. John Wiley & Sons, Chichester.

Gleick, J., 1988. *Chaos. Making a new science*. William Heinemann, London.

Haggett, P. & Chorley, R.J., 1967. Models, paradigms and the new geography. In: R.J. Chorley & P. Haggett (eds) *Models in Geography*. Methuen, London, pp. 19–42.

Haines-Young, R. & Petch, J., 1986. *Physical Geography: Its Nature and Methods*. Paul Chapman Publishing, London.

Harvey, D., 1969. *Explanation in Geography*. Edward Arnold, London.

Henderson-Sellers, A. & McGuffie, K., 1987. *A Climate Modelling Primer*. John Wiley & Sons, Chichester.

IPCC, 1990. *Climatic Change. The I.P.C.C. Scientific Assessment*. J.T. Houghton, G.J. Jenkins & J.J. Ephraums (eds), Cambridge University Press, Cambridge.

Kerr, R., 1989. Hansen vs the World on the Greenhouse effect. *Science*, **244**, 1041–1043.

Kirkby, M.J., Naden, P.S., Burt, T.P. & Butcher, D.P., 1987. *Computer Simulation in Physical Geography*. John Wiley & Sons, Chichester.

Macmillan, B. (ed.), 1989. *Remodelling Geography*. Blackwell, Oxford.

Radford, P.J., 1991. Ecological modelling on personal computers. In: D.G. Farmer & M.J. Rycroft (eds) *Computer Modelling in the Environmental Sciences*, Clarendon Press, Oxford, pp. 309–324.

Thornes, J., 1989. Geomorphology and grass roots models. In: B. Macmillan (ed.) *Remodelling Geography*. Blackwell, Oxford, pp. 3–21.

Part II
A PRACTICAL GUIDE TO COMPUTER MODELLING

3
Making a Paper Model

INTRODUCTION

The previous chapters have introduced environmental systems and the different types of models which are available for environmental work, and outlined some of the concepts involved within each type. This chapter, and indeed the whole of this section of the book, is concerned with translating these fairly abstract concepts into reality. We are concerned here with the construction of mathematical models and particularly with the use of microcomputers in environmental modelling.

The chapter begins with an explanation of the differences between the two principal types of mathematical model: the analytical model and the numerical model. We then consider our first system in detail and choose the hydrological problem of finding the discharge of a stream in Catchwater Catchment from the rainfall records and a knowledge of the various overland and subsurface flow paths. We shall see that an analytical solution to the problem is quite possible, but that the result is restrictive and so, finally, we turn to the idea of numerical modelling. The chapter explains how to use pencil and paper to construct a quite realistic numerical model of the local throughflow paths.

ANALYTICAL AND NUMERICAL MODELLING

Environmental models may be divided into three groups: first there are descriptive models involving sentences, diagrams or maps; second there are empirical models involving the collection and plotting of data; and third there are theoretical models involving the formal statement of processes which link the parameters of interest. We have also seen that descriptive modelling, although a necessary precursor to a more sophisticated analysis of a problem, generally offers little explanation and hardly any predictive capability. The empirical models involve the formulation of some statistical relationship between the parameters of interest and, provided that the original data were half decent, such statistical relationships offer a predictive capability. That is to say an estimate can be made of the output of a system over a range which is different from that covered by the original data. However, empirical models do not attempt to offer an explanation in terms of the operative processes. These have been referred to (see Chapter 2) as 'black box' models. Theoretical models are different again and begin with a formal proposal about the operative processes, and then construct some type of mathematical formula based upon these processes. It is clear, therefore, that provided the results bear some relationship to reality, theoretical models offer both a predictive capability and an explanation which is inherent in the proposed processes.

In this book we are primarily concerned with theoretical modelling and the interested reader is referred to other texts for the principles of statistical analysis and the construction of empirical models. We concentrate on theoretical modelling and argue that the techniques involved in theoretical modelling can be divided into *analytical* and *numerical* procedures. In order to understand the differences, take the example of the Catchwater Catchment illustrated by Figures 3.1 and 3.2. Catchwater Catchment is a small area of Holderness in East Yorkshire. Holderness is bounded to the east by the North Sea, to the south by the River Humber and to the west and north by the Cretaceous chalk hills of the Yorkshire Wolds. The catchment is on the eastern side of Holderness, south of Hornsea, where the bowl-like plain rises imperceptibly towards the coast. The catchment has an area of a little over 15 km^2 in Pleistocene drift deposits and is

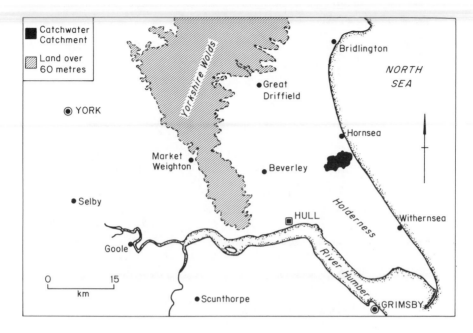

Figure 3.1 *Location of Catchwater Catchment in Holderness: the first objective is to formulate a model of water movements within the area*

fully described in Ward (1967) and Bonnel (1978). The object of the exercise is to determine the channel discharge at the gauging station for a given rainfall. Now, the 'movement of water in the land phase of the hydrological cycle is a complex process' (Shaw, 1983, p.361) and yet it is a major requirement of hydrologists, geomorphologists and indeed the engineers and environmentalists who must manage water to answer the question of rainfall–discharge relationships. For simplification, therefore, the hydrology of the drainage basin, from precipitation through to a stream discharge at the lowest outfall, can be conceived of as a series of interlinked processes and storages, and there have been numerous attempts to construct models of these various elements or of the whole system (e.g. the review by Fleming, 1975 or Ward & Robinson, 1990). Here one of the simpler, earlier models (Dawdy & O'Donnell, 1965) is used to illustrate the principles involved. The model (Figure 3.3) is built around two storages. First

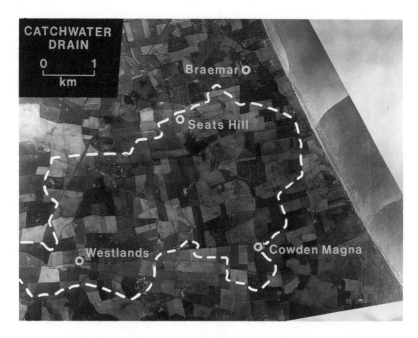

Figure 3.2 *Aerial photograph of the Catchwater Catchment in eastern Holderness showing catchment boundaries and the gauging station*

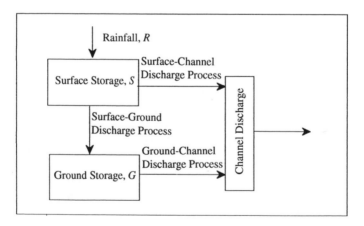

Figure 3.3 *Schematic representation of the Catchwater Catchment system showing input from rainfall, three water transfer processes and the output in terms of the channel discharge*

there is the *surface storage*, which has a content S m^3 of water and within which rainfall is detained on the surface or intercepted by vegetation. Second, there is the subsurface ground water storage, which contains G m^3 of water and for the present purposes will include water in the unsaturated soil layers as well as within the deeper saturated zone beneath the water table.

Let us assume that the excess rainfall, R (i.e. actual rainfall less evaporation losses), enters the surface storage, S, and that some of this, the surface–channel discharge Q_{SC} enters the channel whilst some, the surface–ground discharge Q_{SG} enters the ground where it is added to the ground storage, G. Again, some of the water in the ground storage, the ground–channel discharge S_{GC}, enters the channel, so that the resulting channel discharge is simply the sum of inputs from the surface and from the ground.

This simple model of the system thus involves three processes in addition to the straightforward accounting for water volumes. The theoretical modeller would propose that, for example, the processes involved are:

- *Process 1. Surface–channel discharge process.* The larger the volume of water on the surface, then the greater will be the discharge into the channel. For example:

  Surface–channel discharge = constant$_1$ × surface storage   (3.1)

- *Process 2. Surface–ground discharge process.* The larger the volume of water on the surface, then the greater will be the discharge into the ground. For example:

  Surface–ground discharge = constant$_2$ × surface storage   (3.2)

- *Process 3. Ground–channel discharge process.* The larger the volume of water in the ground, then the greater will be the discharge into the channel. For example:

  Ground–channel discharge = constant$_3$ × ground storage   (3.3)

The result would be four equations (the relationship between surface storage and rainfall which includes the surface area of the catchment, together with the three relationships outlined above between surface–channel discharge and surface–ground discharge and surface storage and between ground–channel discharge and ground storage) along with the summation of the surface–channel discharge and ground–channel discharge to calculate the total channel discharge. In addition there are now at least four system parameters (including the three constants and the catchment area). The modeller would write down the equations and, using a little algebra, would be able to solve them by making certain assumptions about the values of the different constants and the maximum. This single equation represents the *analytical solution* to the problem, but it is clear that there are very many roots (i.e. combinations of parameters) which offer a working solution. It is just this complexity which has moved the environmental scientist beyond such analytical solutions towards numerical methods and the construction of numerical models. For one thing the mathematics is rather simpler: the above example represents a very simplified system and environmental systems are, almost by definition, very complex calling for many, sometimes many tens and occasionally many hundreds, of equations to be solved simultaneously and this is rarely possible if any degree of realism is required. Numerical solutions almost invariably involve computers and do not actually solve the equations simultaneously, but rather leave them in their original form and let the computer search for possible solutions. We shall now progress to a numerical solution to the Catchwater Catchment problem.

THE CATCHWATER CATCHMENT MODEL

The Catchwater Catchment system consists of rain falling over a particular area, being stored or transported through a series of subsystems and the result is that some, but by no means all, of the rainwater exits through the flume channel. This is a complex system because, even at a trivial level, there is an interaction between many of the variables and it is almost impossible to appreciate the relative or absolute magnitudes of these various flow paths with any degree of certainty. The full system is even more complex when such processes as evaporation, evapotranspiration, or the various auxiliary water storages are considered. The present model, therefore, will be a simplification but nevertheless the principles of modelling should become apparent.

We shall consider only one input parameter, three fixed parameters which essentially describe the state of the system before we start work, and three processes, and aim for a single output parameter. It is firstly necessary to clearly state the problem, and this is shown by the diagram in Figure 3.3. We shall assume that the rainfall (in units of cubic metres per square metre of catchment surface per month) falls equally throughout the catchment. Thus the product of the rainfall and the surface area of the catchment defines the volume of water entering the system in the month. We shall assume that a certain volume of water was present in the surface store at the beginning of the month, and that this is supplemented by the rainfall, and that a representative figure for the surface storage will be determined by adding the existing water to one half of that falling over the month. This water is transmitted by the first two of the three processes detailed above (equations (3.1) and (3.2)) either directly into the channel or indirectly into the ground store. We shall further assume that a certain amount of water was already present in the ground store and, again, this is augmented by one half of the volume arriving from the surface to give a mean ground store value for the month. Finally, we shall assume that a proportion of the mean ground store volume flows into the channel in accordance with the third process (equation (3.3)) and that the mean channel discharge is then given by the sum of the surface and ground store inputs. This value will be in units of cubic metres per month, and is converted into units of cubic metres per second (cumecs) for comparison with field measurements. There are three process equations plus a further five calculations required to describe the system. It would, of course, be possible to solve all eight equations simultaneously and the result would be:

$$\text{Channel discharge} = \frac{C_1(S + 0.5(A.R)) + C_3(G + 0.5.C_2(S + 0.5(A.R)))}{\text{days} \times \text{seconds per day}}$$

where, to recap: C_1, C_2 and C_3 are coefficients in the processes,
                          R is the rainfall,
                          A is the surface area of the catchment,
                          S is the initial contents of the surface store, and
                          G is the initial contents of the ground store.

It is apparent, however, that this is hardly an elegant solution, and we shall see that a more useful model is achieved by choosing the numerical route. In the following section this numerical solution is implemented on a piece of paper, and in the next chapter the model is tranferred to the microcomputer.

CONSTRUCTING A PAPER MODEL OF THE CATCHWATER CATCHMENT

The simplest and always the first step in the construction of a computer model is to take a pencil and a clean piece of lined paper. For the present, rule three columns as shown in Figure 3.4, and lay out the rows to correspond to the various parameters. Use Columns A and B simply for text where column B is for any equations which are required and use column C for the actual numbers, i.e. for the values of the inputs and parameters and for the results of the calculations. Write in the appropriate text and complete the eight equations in column B (Figure 3.4) and then write in the corresponding seven inputs and parameter values (Figure 3.5) using, in this example:

- Initial surface storage   C2   0
- Rainfall   C3   0.128
- Catchment area   C4   15
- Constant 1   C7   0.9
- Constant 2   C8   0.6
- Constant 3   C9   0.3
- Initial ground storage   C14   0

	Column A	Column B	Column C
1			January
2	Initial surface storage cu m		
3	Rainfall cu m/sq m/month		
4	Area sq km		
5	Total water input cu m/month	C5=C3*C4*1000000	
6	Mean surface storage cu m	C6=	
7	constant 1		
8	constant 2		
9	constant 3		
10			
11	Surface to channel cu m/month	C11=	
12	Surface to ground cu m/month	C12=	
13			
14	Initial ground storage cu m/month		
15	Mean ground storage cu m/month	C15=	
16	Ground to channel cu m/month	C16=	
17			
18	Total channel cu m/month	C18=	
19	Mean channel discharge cumecs	C19=	

Figure 3.4 *Inputting the text into the paper model*

	Column A	Column B	Column C
1			January
2	Initial surface storage cu m		0
3	Rainfall cu m/sq m/month		0.128
4	Area sq km		15
5	Total water input cu m/month	C5=C3*C4*1000000	
6	Mean surface storage cu m	C6=C2+0.5*C5	
7	constant 1		0.9
8	constant 2		0.6
9	constant 3		0.3
10			
11	Surface to channel cu m/month	C11=C7*C6	
12	Surface to ground cu m/month	C12=C8*C6	
13			
14	Initial ground storage cu m/month		0
15	Mean ground storage cu m/month	C15=C14+0.5*C12	
16	Ground to channel cu m/month	C16=C9*C15	
17			
18	Total channel cu m/month	C18=C11+C16	
19	Mean channel discharge cumecs	C19=C18/(31*86400)	

Figure 3.5 *Inputting the numbers into the paper model*

Finally use a calculator to work out the results of the various equations and hence the mean channel discharge (Figure 3.6). There are a number of aspects of the paper model which should be noted at this stage. Firstly it is convenient to use the column letters and row numbers to designate particular parameter values. Thus the first equation, that the total water input is equal to the product of the rainfall and the surface area of the catchment, is written as

$$C5 = C3 * C4 * 1000000$$

because the rainfall value is contained in column C row 3 and the area of the catchment is in C4. The one million converts the area of the catchment in square kilometres to square metres. We shall use this type of designation in later chapters and it is conventional to always quote the column letter before the row number. Second, note that it takes quite a long time to calculate and to write down the

	Column A	Column B	Column C
1			January
2	Initial surface storage cu m		0
3	Rainfall cu m/sq m/month		0.128
4	Area sq km		15
5	Total water input cu m/month	C5=C3*C4*1000000	1,920,000
6	Mean surface storage cu m	C6=C2+0.5*C5	960,000
7	Constant 1		0.9
8	Constant 2		0.6
9	Constant 3		0.3
10			
11	Surface to channel cu m/month	C11=C7*C6	864,000
12	Surface to ground cu m/month	C12=C8*C6	576,000
13			
14	Initial ground storage cu m/month		0
15	Mean ground storage cu m/month	C15=C14+0.5*C12	288,000
16	Ground to channel cu m/month	C16=C9*C15	86,400
17			
18	Total channel cu m/month	C18=C11+C16	950,400
19	Mean channel discharge cumecs	C19=C18/(31*86400)	0.35

Figure 3.6 Completing the calculations in the paper model

results. At this stage we are occupying a half-way house somewhere between the cumbersome analytical solution described earlier, and the numerical solutions described later. Complete this chapter by trying the examples below.

EXAMPLES

1. Write out four copies of the paper model as in Figure 3.4 with the following values:

Rainfall	C3	0.128
Catchment area	C4	15
Constant 1	C7	0.9
Constant 2	C8	0.6
Constant 3	C9	0.3

   and complete column B after the equals signs (i.e. check the formulae in Figure 3.5.)
2. Complete column C and calculate the discharge in cumecs (i.e. check Figure 3.6) with the following assumptions: C2 = 0 and C14 = 0.
3. Repeat as in Example 3.2 but assume that: C2 = 10 000.
4. Repeat as in Example 3.3 but assume that: C14 = 250 000.
5. Repeat as in Example 3.4 but assume that: C3 = 0.1.
6. Make a note of how long it has taken you to complete this exercise.

REFERENCES

Bonell, M., 1978. An evaluation of shallow groundwater movement in a small boulder clay catchment in Holderness. *Miscellaneous Series* No.18. University of Hull.

Dawdy, D.R. & O'Donnell, T., 1965. Mathematical models of catchment behaviour. *Proc. Am. Soc. Civ. Eng.*, HY4, **91**, 123–137.

Fleming, G., 1975. *Computer Simulation Techniques in Hydrology*. Elsevier, Amsterdam.

Shaw, E.M., 1983. *Hydrology in Practice*. Van Nostrand Reinhold. Wokingham, England.

Ward, R.C., 1967. Design of catchment experiments for hydrological studies. *Geog. Jnl.*, **133**, 495–502.

Ward, R.C. & Robinson, M, 1990. *Principles of Hydrology*, 3rd edn. McGraw-Hill, Maidenhead.

4
Making a Computer Model

INTRODUCTION

The previous chapter was concerned with deriving the controlling equations for a simple model and with laying that model out on a sheet of paper. The choice of model, and of layout style was deliberate because the present chapter is concerned with transferring the model to the Excel spreadsheet programming environment. We shall see that no changes in layout and only minimal changes in the equations are required to proceed from the rather cumbersome paper model to the slick computer model. This chapter explains how to start-up an Excel file, and how to enter the model.

STARTING AN EXCEL FILE

The PC or Macintosh should be switched on, and Excel should be available. Then click on the Model.blank icon (Figure 4.1), which opens an empty Excel file (Figure 4.2). If the Model.blank file is not available then refer to Appendix I for details of its construction. The screen display now has 14 different elements, and the user should read the following and become familiar which the location of each one.

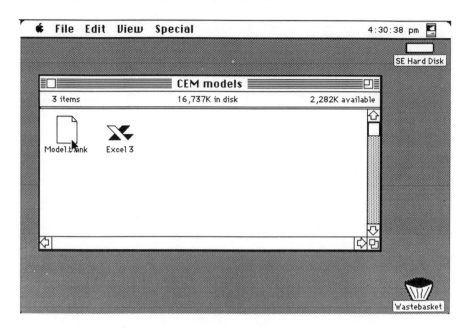

Figure 4.1   Opening the Excel file Model.blank by clicking the icon

1. *Menu Bar*. The menu bar contains the Apple logo, plus file, edit, formula, format, data, options, macro and window menus. Click on one of these and check that the menu is displayed. For example, click on 'format' and number, alignment, style, border, cell protection and column width options are displayed. Click on one of these and check that the sub-menu is displayed. For example, click on 'alignment' and general, left, centre, right and fill options are displayed together with the OK and cancel buttons. Click on 'cancel' without changing the settings and return to the full display as shown in Figure 4.2.
2. *Document name*. The document name is presently Model.blank but can be changed when saving the work through the file option.
3. *Cells*. Cells are the individual elements which comprise the worksheet. They are presently rectangular, but their width and depth can be altered with the format and options menus.
4. *Column Headings*. As in the previous chapter, column headings refer to the vertical lines of cells called columns. They are

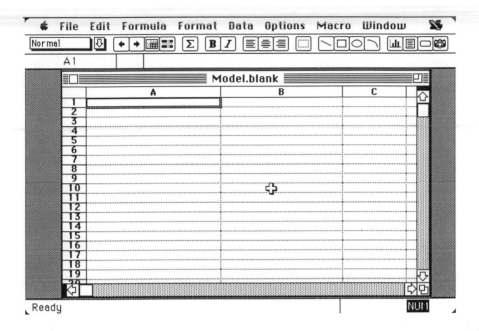

Figure 4.2 *The screen display at the beginning of the session*

headed A, B, C, etc. The columns are labelled from the left-hand side of the worksheet with A to Z and then BA to BZ and so on to column IV. There are thus up to 256 columns on an Excel worksheet.

5. *Row Headings*. As in the previous chapter, row headings refer to the horizontal lines of cells called rows. Rows are labelled from the top of the worksheet and from 1 to 16 384. Cells are always referred to by their column and then by their row, thus A1 or B17, etc.

6. *Reference to active cell*. The active cell presently contains A1.

7. *Active cell*. The active cell is highlighted by a box and corresponds to the reference to the active cell above.

8. *Pointer*. The pointer is presently a + sign when on the worksheet but reverts to an arrow when elsewhere on the screen display. The pointer is moved with the mouse. Point to cell B7 and click the mouse. Cell B7 now becomes the active cell and the reference to the active cell on the display changes accordingly.

9. *Scroll bars*. Scroll bars lie along the bottom and right-hand margins of the worksheets, and are operated in the same way as with other Macintosh applications through the scroll arrows and scroll boxes.

10. *Scroll arrows*. Scroll arrows lie at either end of the scroll bars and, when clicked, move the display a short distance across the worksheet.

11. *Scroll boxes*. Scroll boxes lie on both of the scroll bars and move the display rapidly across the worksheet.

12. *Size box*. The size box lies at the bottom right-hand corner of the worksheet and is used to change the area displayed on the screen. It is used in the same way as in other Macintosh applications.

13. *Tool bar*. The tool bar lies beneath the menu bar and contains (from the left) the two styles boxes, followed by four outlining buttons, followed by the summation button, five formatting buttons (bold, italic, left, centre and right justify), a selection tool, and then a set of four drawing tools and finally the chart, text box, button and camera tools. The function of these various elements is described in later sections, it is sufficient to realise that, in general, the 'buttons' duplicate commands within the pull-down menus whilst the 'tools' are used to enhance the screen displays.

14. *Formula bar*. The formula bar lies beneath the tool bar and is used for editing the contents of each cell as described in the following section.

CELL TYPES

The computer model operates by defining the relationship between cells in the worksheet. These relationships are called formulae in Excel, and actually only occupy a small part of the completed worksheet. Additionally the parameters which are used as input into each formula occupy some of the cells, whilst the outputs occupy other cells. Excel also allows for a very liberal sprinkling of information which is not central to the running of the model, but which is useful in understanding how the model works. This type of information is usually in the form of words or phrases and is ignored by the model when it is performing calculations. There are, therefore, only three types of information which need to be entered into the cells on the worksheet during the construction of the model. These are text, numbers and formulae, and each will be detailed in the following sections.

Entering Text

In order to enter anything into the worksheet, the appropriate cell must firstly be selected and identified as the 'active cell'. Do this as before by moving the pointer and clicking the required cell. It should be highlighted by the box and its column and row will appear in the reference to the active cell above the worksheet. For example, begin to construct the hydrological model of Catchwater Catchment which was produced in the previous chapter by:

1. *Selecting cells.* Move the pointer to A2 and click once. The cell will be highlighted with a box and A2 will be displayed as the reference to active cell on the screen.
2. *Editing text.* Type 'Initial surface storage cu m' which appears in the formula bar on the screen display. Any mistakes during typing can be rectified with the backspace key or by using the mouse to reposition the cursor on the formula bar and to insert additional letters.
3. *Entering text.* There are two ways in which text in the formula bar can actually be entered into the cell on the worksheet:
   (a) Press the 'return' key when the cursor is at the end of the text on the formula bar.
   (b) Click on the tick which is displayed to the left of the text on the formula bar.
   It is simpler to use the second method to enter 'Initial surface storage cu m' into cell A2.
4. *Check the entry.* Ensure that the text which you want appears in the correct cell on the worksheet. Any errors can be corrected by repeating the first three steps.
5. *Complete text entry.* Complete the entry of text into the worksheet

by selecting, editing and entering text as above for the remainder of the words in column A as shown in Figure 4.3. In addition column B will be used simply to contain further text which explains the computer model, as it was used in the previous chapter to explain the paper model. Therefore enter 'C5=C3*C4*1000000' into cell B5 and so on until column B is also complete as shown in Figure 4.3.

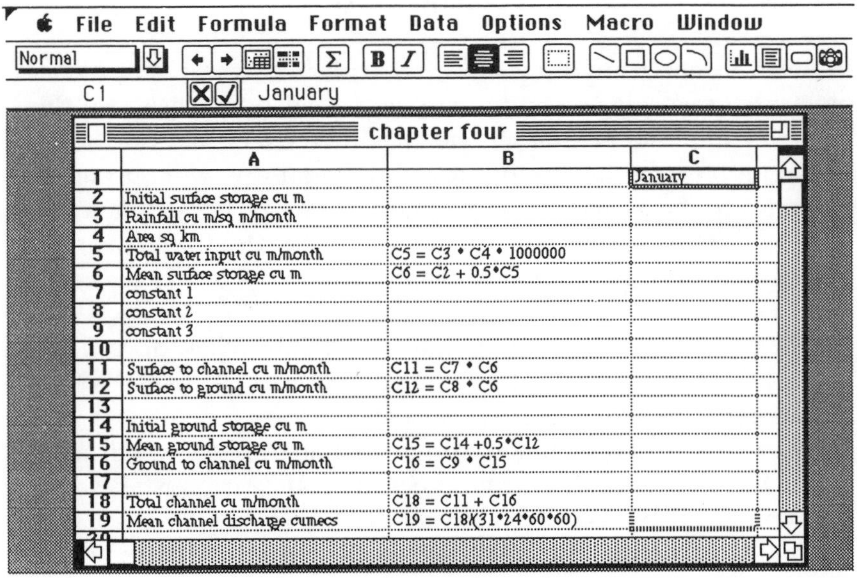

Figure 4.3   *Entering text into the catchment model*

Entering Numbers

The paper model which was developed in Chapter 3 included certain inputs and parameters which were assumed to be constant such as the rainfall, catchment area and so on. These numbers can be entered into the computer model in the same way as text by:

1. *Selecting cells.* Move the pointer to cell C3 and click once.
2. *Editing numbers.* Type '0.128' (the rainfall) into the formula bar.

3. *Entering numbers.* Click on the tick to enter 0.128 into cell C3.
4. *Check the entry.* Ensure that 0.128 has appeared in cell C3.
5. *Complete number entry.* Complete the entry of numbers into the worksheet by selecting, editing and entering the constants into column C as shown in Figure 4.4.

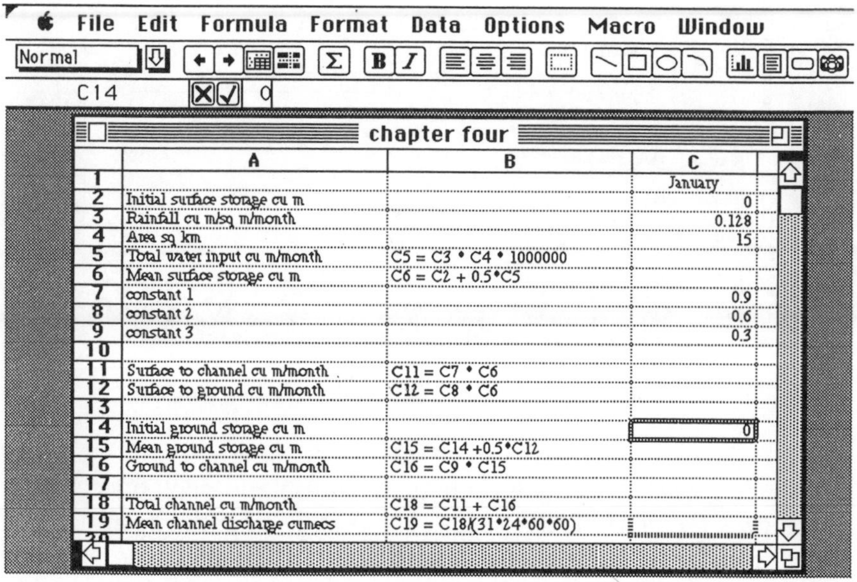

Figure 4.4 *Entering numbers into the catchment model*

Entering Formulae

The third and final type of information which is entered into the worksheet is the formulae which actually represent the model. These are, for example, the requirement that cell C5, the total water input, should be equal to the product of cells C3 and C4 the rainfall and the surface area of the catchment multiplied by one million to convert to units of cubic metres. There are in fact only eight formulae in the catchment model and these are to be entered in cells C5, C6, C12, C13, C15, C16, C18 and C19. They are entered as shown in Figure 4.5 by:

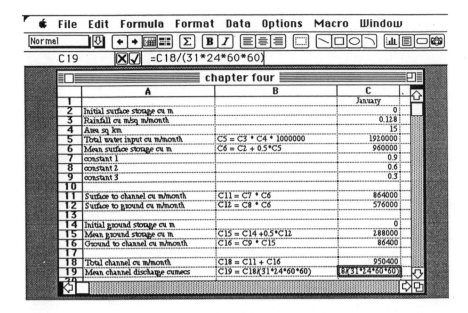

Figure 4.5 *Entering formulae into the catchment model*

1. *Selecting cells.* Move the pointer to cell C5 and click once.
2. *Editing formulae.* Type '=C3*C4*1000000' into the formula bar. Notice that it is not necessary to specify C5=, because once the chosen cell has become active when entering the formula, the result of the calculation is placed in that cell. This is another way of saying that all formulae begin with the equals sign, and this is how Excel recognises a formula and distinguishes it from text or numbers.
3. *Entering formulae.* Click on the tick to enter '=C3*C4*1000000' into cell C5.
4. *Check the entry.* As soon as a new formula is entered onto the worksheet the result of the calculation appears in the cell. In the present case, C5 contains the product of C3 and C4 multiplied by one million, that is 1 920 000.
5. *Complete formulae entry.* The other seven formulae are entered onto the worksheet by selecting, editing and entering the formulae which were derived in the previous chapter to complete the model. These have already been written in text form in

column B, and should be reproduced beginning with the equals sign in the appropriate cells in column C. Check that the following have been correctly entered by selecting each cell in turn and reading the formula which it contains in the formula bar:

Cell	Formula
C5	=C3 * C4 * 1000000
C6	=C2 + 0.5 * C5
C11	=C7 * C6
C12	=C8 * C6
C15	=C14 + 0.5 * C12
C16	=C9 * C15
C18	=C11 + C16
C19	=C18/(31*24*60*60)

Note that a multiplication sign is represented by an '*' in Excel as in cell C5, and that the multiplication sign must be included in front of any brackets. Also note that the divide sign is represented by '/' in Excel as in cell C19.

RUNNING THE MODEL

The model is now complete, and all of the formulae are in fact working. The model represents the first case examined with the full paper model at the end of the previous chapter. The result is shown in Figure 4.6, and is predicting a mean stream discharge for January of 0.35 cumecs, a not unreasonable value.

This simple computer model has been used to illustrate the construction of any model in Excel, and it should now be used to examine any number of 'what if' scenarios. The example below is used to illustrate some possibilities, but the main objective is to show that, once the computer model has been constructed and checked, then it becomes a very trivial task to change any of the input parameters, or indeed any of the formulae to examine other possibilities. This theme is developed in the following chapters where this and other models are coded and operated. The tasks

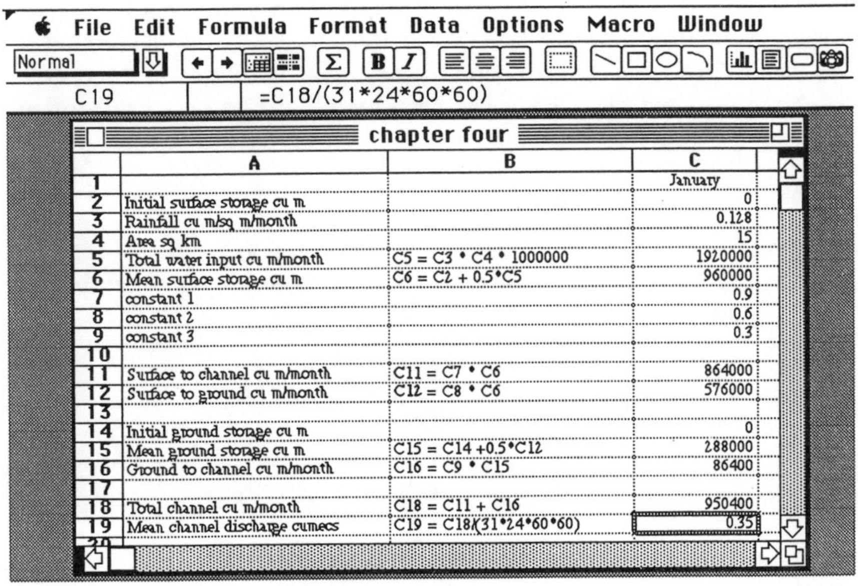

File   Edit   Formula   Format   Data   Options   Macro   Window

Normal				

C19 =C18/(31*24*60*60)

chapter four

	A	B	C
1			January
2	Initial surface storage cu m		0
3	Rainfall cu m/sq m/month		0.128
4	Area sq km		15
5	Total water input cu m/month	C5 = C3 * C4 * 1000000	1920000
6	Mean surface storage cu m	C6 = C2 + 0.5*C5	960000
7	constant 1		0.9
8	constant 2		0.6
9	constant 3		0.3
10			
11	Surface to channel cu m/month	C11 = C7 * C6	864000
12	Surface to ground cu m/month	C12 = C8 * C6	576000
13			
14	Initial ground storage cu m		0
15	Mean ground storage cu m	C15 = C14 +0.5*C12	288000
16	Ground to channel cu m/month	C16 = C9 * C15	86400
17			
18	Total channel cu m/month	C18 = C11 + C16	950400
19	Mean channel discharge cumecs	C19 = C18/(31*24*60*60)	0.35

Figure 4.6 *The completed catchment model*

used here have only analysed the smallest part of the powerful programming and display capabilities of Excel, and certain other aspects will be introduced at a later stage.

EXAMPLES

1. Repeat the examples in the previous chapter for each of the four cases and check that the same mean stream discharges are achieved.

2. Using case three from the paper model in the previous chapter and by varying only the initial value in the surface store, determine the value that would be required to produce a mean stream discharge of 0.5 cumecs.

5
Environmental Models

INTRODUCTION

The previous chapter was concerned with transferring the catchment model to the Excel spreadsheet. The result was a single column of six formulae which calculated the mean discharge for one month in a simplified hydrological system. Geography is concerned with systems which range from such simple examples to the very complex (see Chapter 2). In particular, geographers are concerned with modelling the behaviour of environmental systems (see Chapter 1) as they change through time (evolutionary models) and through space, i.e. across the surface of the Earth. This chapter is, therefore, concerned with extending the catchment model from one month to six months and then from one catchment to six catchments. We shall see that only small changes in layout and minimal changes in the equations are required to proceed from the rather limited computer model of January to a more sophisticated geographical model. This chapter explains how to extend an Excel file, and how to format and duplicate cell contents and formulae.

EXTENDING THE WORKSHEET

The worksheet that was developed in the previous chapter consisted of only three columns, and only one of these contained formulae for the model. In this section the worksheet is extended to cover additional columns.

1. *Additional columns.* Open the model from the previous chapter and return to the case 1 settings as was shown in Figure 3.5. Type 'February' into cell D1, 'March' into cell E1, 'April' into F1 and so on until June has been entered into H1 using the scroll arrows on the horizontal scroll bar at the bottom of the screen to move the display to the right. The result is shown in Figure 5.1. There are two problems. Firstly we cannot now see the whole of the worksheet on the screen, and secondly the word 'January' is centralised in the column, but 'February', 'March' and so on are on the left-hand side. We correct the size of the screen display in the following section and then progress to align the text more uniformly in the columns.

2. *Size of the screen display.* Move the pointer to the document name bar, hold down the mouse button and drag the worksheet to

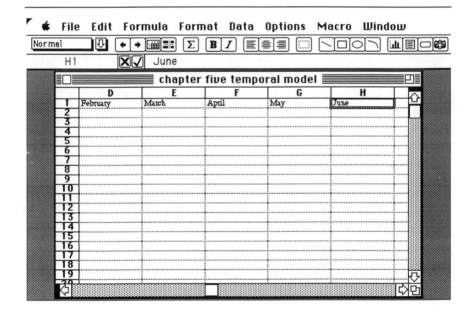

Figure 5.1 *Adding additional columns to the budget model*

the top left-hand side of the display screen. Now use the size box to extend the right-hand side of the worksheet to the right-hand side of the screen. Ensure that the 19th row remains visible. The screen is now larger but all of the cells are still not visible.

HIGHLIGHTING MULTIPLE CELLS

The technique of making a single cell active was described in the previous chapter. When, however, similar changes are to be made to more than one cell it is useful to make larger portions of the worksheet active at one time. This is called highlighting because it has the effect of reversing the colour of the display on the active cells.

Removing Column B

1. Column B was employed in the paper model to explain the use of formulae in column C and is now redundant and can be removed. Move the pointer to the column header cell in column B and press the mouse. This highlights all of column B (Figure 5.2).
2. Any action taken will now apply to all of the highlighted cells. Check by clicking on D8 and then returning to highlight all of column B.
3. Open the edit menu and click the delete option (Figure 5.2). The old contents of column B are removed from the spreadsheet and all of the remaining columns are moved to the left.

Highlighting Multiple Columns

1. Click on the new B column header cell to again highlight column B, but do not release the mouse button.
2. Keep the button depressed and move the pointer to the right through C, D and so on until G, the June column, is highlighted; then release the button.

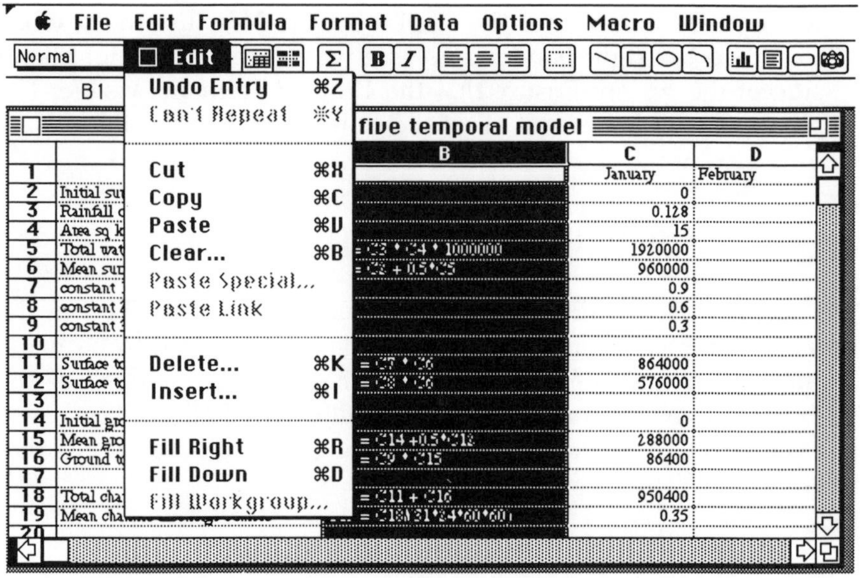

Figure 5.2 *Highlighting and deleting a column from the worksheet*

3. Every column from B to G should now be highlighted and any action applies to all of these cells. We shall now improve the appearance of all of these cells.

FORMATTING MULTIPLE CELLS

The layout and appearance of cells on the worksheet is called the format, and the width, justification and number type can each be formatted from the menu bar:

Column Width

1. With columns B to G highlighted as above, click on the Format menu in the menu bar and, without releasing the mouse button, drag down to highlight the column width option. Release the mouse button and the column width sub-menu is displayed as in Figure 5.3.

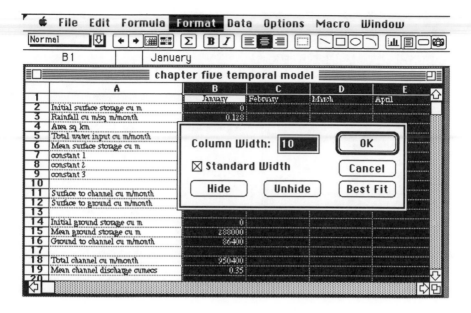

Figure 5.3 *Choosing the column width option in the format menu*

2. The cells are presently 10 units wide, and this additional width is preventing the full worksheet from being displayed simultaneously. Therefore click on the column width box in the sub-menu and delete the 10, and type in 6.
3. Click 'OK.'
4. The spreadsheet is re-formatted automatically with a column width of six units being applied to all of the highlighted cells.

Column Justification

1. The column headings February to June are, however, still on the left-hand side of the cells (and so would be any figures entered into these columns) whilst January is in the centre as required. Correct this fault by highlighting the February cell (C1) and dragging the pointer to the right whilst keeping the mouse button depressed to multiple highlight cells C1 to G1.

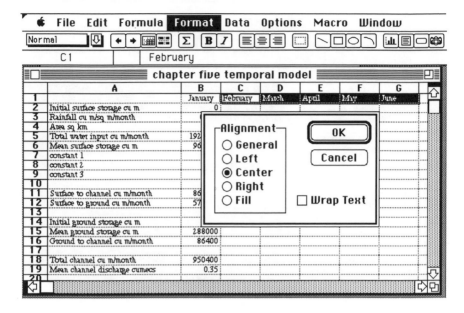

Figure 5.4 *Choosing the alignment option to right justify the cells*

2. Return to the format menu and this time choose the alignment sub-menu.
3. Click on the 'centre' pip box by pointing to it and pressing the mouse button once (Figure 5.4).
4. Click 'OK.'
5. The months are now properly aligned in the centre of the cells.
6. Note that an alternative to using the alignment sub-menu is to highlight cells C1 to G1 as above and then to click on the centre button on the tool bar. This button is in fact displayed in bold on the screen display after completing the steps above which indicates that the highlighted cells have centre alignment applied.

Number Type

1. The number format for the columns is still undefined, and we require integers for most of the numbers, two decimal places

for the three constants in rows 7–9 and three decimal places for the rainfall and for the mean channel discharge in cumecs. It is easiest to apply integer format to all of the numbers and then to correct the decimal places for the constants and rainfall and discharge. Begin by multiple highlighting cells B2 to G19.

2. Return to the format menu, this time choosing the numbers sub-menu.
3. Choose the 0 option which signifies integers as in Figure 5.5.

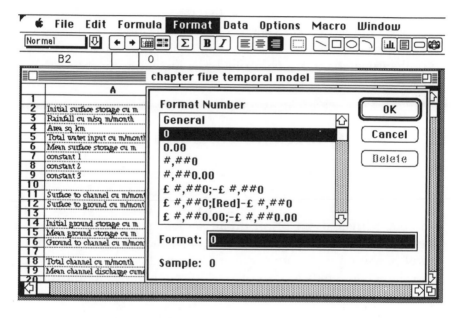

Figure 5.5 *Choosing the number option in the format menu*

4. Click 'OK.'
5. All figures in columns B to G will now be represented in integer format. Whilst all these cells are highlighted, choose the right alignment option in the format menu to keep the figures to one side of the cells.
6. The rows which represent the constants are, however, now formatted as integers and we wish them to be two decimal

places. Correct this fault by highlighting B7 and, keeping the mouse button depressed, drag the cursor across to G9 and then release so that all the constant cells are highlighted.

7. Return to the format menu and choose the number sub-menu as before.

8. Choose the 0.00 option which represents two decimal places.

9. Click 'OK.'

10. Finally, set rows 3 and 19 to three decimal places by returning to the number sub-menu and typing '0.000', followed by clicking 'OK.' The number format across the spreadsheet is now as we would wish, and we discover that the January mean discharge is 0.355 rather than 0.35 cumecs. This illustrates that, although the displays are fixed to the precision chosen in the format menu, the calculations are performed to a greater accuracy.

MODELLING THROUGH TIME

The extension of the catchment model from one month to six months is an example of a time dependent or evolutionary model. This type of problem is common in geography when the development of, for example, a river basin or an urban settlement may be of interest. The model is said to be time dependent which means that one (or more) of the input parameters varies through time and the model is designed to predict the outcome through time. In the present example we shall make only two assumptions:

1. Firstly we shall assume that the only time-dependent parameter in the model is the rainfall in each month.

2. Secondly we shall assume that the mean surface storage in each month becomes the initial surface storage in the following month and, similarly, the mean ground storage in each month becomes the initial ground storage in the following month.

The following section explains how to enter these changes to complete the temporal model. The changes involve either entering new cells or duplicating existing ones.

ENTERING NEW CELLS

1. Enter 0.153, 0.200, 0.147, 0.095 and 0.095 into cells C3 to G3 respectively to represent the rainfall in each month.
2. Enter the following formulae to carry forward the water storages from the previous month (Figure 5.6):

Surface storage		*Ground storage*	
Cell C2	=B6	Cell C14	=B15
Cell D2	=C6	Cell D14	=C15
Cell E2	=D6	Cell E14	=D15
Cell F2	=E6	Cell F14	=E15
Cell G2	=F6	Cell G14	=F15

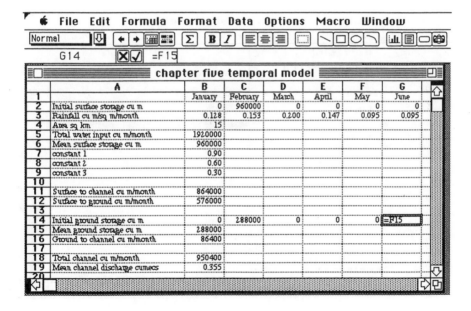

Figure 5.6 *Entering new formulae into the worksheet*

DUPLICATING CELLS

Inserting these new numbers and formulae into the worksheet is a necessary but fairly repetitive stage in the development of the time-dependent model. Fortunately Excel facilitates the duplication of cells as follows.

Duplicating Constants

1. The surface area of the catchment is assumed to be constant.
2. Highlight cell B4 and, without releasing the mouse button, drag the cursor to the right to highlight the B4 to G4 row.
3. Open the edit menu and choose the 'fill right' option as shown in Figure 5.7.
4. The number 15 is automatically placed into all of the highlighted cells.

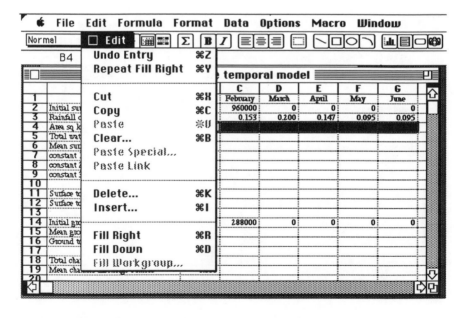

Figure 5.7 *Duplicating numbers and constants in the model*

5. Repeat the process for the other process parameters: constant 1 (B7 to G7), constant 2 (B8 to G8) and constant 3 (B9 to G9).

Duplicating Formulae

There are eight formulae in the January column of the model at present and these are extended to the right in a similar manner to the constants above.

1. Highlight cell B5 which contains '=B3*B4*1000000' as displayed in the menu bar.
2. Highlight B5 to G5 and fill right as before.
3. Now (and this is the clever part) inspect the result by highlighting cell E5 (for example, as shown in Figure 5.8) and you will see that Excel has not only duplicated the formula, but has also changed the formula into the correct form for each column, '=E3*E4*1000000' in column E, and so on.

🍎 File Edit Formula Format Data Options Macro Window

Normal	🔽	← → ▦ ▦ Σ B I ≡ ≡ ≡ ⬚ ◥□○◤ 📊🗐○📷

E5		=E3*E4*1000000

| ▤□ | chapter five temporal model | 🔲 |

	A	B	C	D	E	F	G	
1		January	February	March	April	May	June	⬆
2	Initial surface storage cu m	0	960000	0	0	0	0	
3	Rainfall cu m/sq m/month	0.128	0.153	0.200	0.147	0.095	0.095	
4	Area sq km	15	15	15	15	15	15	
5	Total water input cu m/month	1920000	2295000	3000000	2205000	1425000	1425000	
6	Mean surface storage cu m	960000						
7	constant 1	0.90	0.90	0.90	0.90	0.90	0.90	
8	constant 2	0.60	0.60	0.60	0.60	0.60	0.60	
9	constant 3	0.30	0.30	0.30	0.30	0.30	0.30	
10								
11	Surface to channel cu m/month	864000						
12	Surface to ground cu m/month	576000						
13								
14	Initial ground storage cu m	0	288000	0	0	0	0	
15	Mean ground storage cu m	288000						
16	Ground to channel cu m/month	86400						
17								
18	Total channel cu m/month	950400						
19	Mean channel discharge cumecs	0.355						⬇
20								

Figure 5.8 *Duplicating formulae in the model*

| | File | Edit | Formula | Format | Data | Options | Macro | Window |

Normal   🔲   ← → 🏢 📑   Σ   **B** *I*   ≣ ≣ ≣   ⬚   ⬝⬜⭕⬞   📊 📰 🔲 🌐

A1

═══════════════ chapter five temporal model ═══════════════

	A	B	C	D	E	F	G	
1		January	February	March	April	May	June	
2	Initial surface storage cu m	0	960000	2107500	3607500	4710000	5422500	
3	Rainfall cu m/sq m/month	0.128	0.153	0.200	0.147	0.095	0.095	
4	Area sq km	15	15	15	15	15	15	
5	Total water input cu m/month	1920000	2295000	3000000	2205000	1425000	1425000	
6	Mean surface storage cu m	960000	2107500	3607500	4710000	5422500	6135000	
7	constant 1	0.90	0.90	0.90	0.90	0.90	0.90	
8	constant 2	0.60	0.60	0.60	0.60	0.60	0.60	
9	constant 3	0.30	0.30	0.30	0.30	0.30	0.30	
10								
11	Surface to channel cu m/month	864000	1896750	3246750	4239000	4880250	5521500	
12	Surface to ground cu m/month	576000	1264500	2164500	2826000	3253500	3681000	
13								
14	Initial ground storage cu m	0	288000	920250	2002500	3415500	5042250	
15	Mean ground storage cu m	288000	920250	2002500	3415500	5042250	6882750	
16	Ground to channel cu m/month	86400	276075	600750	1024650	1512675	2064825	
17								
18	Total channel cu m/month	950400	2172825	3847500	5263650	6392925	7586325	
19	Mean channel discharge cumecs	0.355	0.811	1.436	1.965	2.387	2.832	
20								

Figure 5.9   *The completed model for catchment discharge through time*

4. Repeat the procedure for rows 6, 11, 12, 15, 16, 18 and 19 to complete the temporal model for Catchwater Catchment as shown in Figure 5.9.

RUNNING THE TEMPORAL MODEL

The time-dependent catchment model for January to June is now completed and any of the inputs could be altered to examine a wide range of 'what if' scenarios. Some possibilities are presented in the Examples section below, but before attempting them, save the model to disk with an appropriate title (Time?) and consider the analogous type of environmental model which involves the examination of spatially-dependent systems.

MODELLING THROUGH SPACE

A related though different extension of the catchment model from one month at one site to one month at six separate sites is an example of a spatially-dependent geographical model. The type of problem is perhaps even more common in environmental work than the time-dependent models which were outlined in the preceding section because geographers are generally concerned with the distribution of parameters across the surface of the Earth. Examples of such spatially-dependent problems are climatic effects on bio-production, or population densities as a function of distance from some focal point. The model is spatially dependent which means that one (or more) of the parameters varies with distance and the model is designed to predict the outcome through distance. In the present example we shall make only two assumptions:

1. Firstly we shall assume that there are six catchments called Deep South, South, Mid South, Mid North, North and Deep North ranging from a wet tropical climate to a dry, arid climate and that the rainfall decreases linearly from the Deep South to the Deep North.
2. Secondly we shall assume that all of the catchments are identical and have initial surface storage and initial ground storages of a nominal 1000 m^3.

The spatial model which results is shown in Figure 5.10 and the last version of the temporal model is converted to this form by the following procedure:

1. Change the column titles in cells B1 to G1 into Deep South, South, Mid South, Mid North, North and Deep North.
2. Type 1000 into B2 and then highlight cell B2 and drag right to multiple highlight B2 to G2 and then fill right to give 1000 m^3 initial surface storage in all catchments.
3. Type 1000 into B14 and then highlight cell B14 and drag right to multiple highlight B14 to G14 and then fill right to give 1000 m^3 initial ground storage in all catchments
4. Finally make the rainfall reflect the climate by using, for example, 0.5, 0.4, 0.3, 0.2, 0.1 and 0.0 for cells B3 to G3 respectively.

The spatially-dependent model is now completed and should be saved to disk with a different title (Space?). It can now be used for examining a range of 'what if' scenarios.

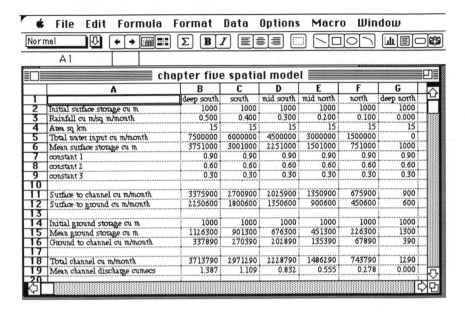

Figure 5.10   *The complete spatial model*

EXAMPLES

1. Complete the time-dependent model as shown above and comment upon the differences between the change in rainfall from January to June (which reaches a minimum in June) in comparison with the change in stream discharge (which does not reach a minimum in June). What is the effect of increasing the initial surface storage volume in January?

2. Complete the spatially-dependent model, and comment upon the differences between the change in rainfall (which is linear with distance) from the south to the north in comparison with the change in stream discharge. What is the effect of increasing the porosity of the catchments (i.e. decrease the three constants to inhibit throughflow)?

6
Presenting the Results

INTRODUCTION

The previous Chapter was concerned with extending the catchment model within the Excel spreadsheet. The result was firstly a temporal model with six columns of six formulae which calculated a mean channel discharge rate for the months from January to June for the simplified catchment system. Secondly a spatial model was developed showing the discharge variations for six catchments from the Deep South to the Deep North. The output from the model was, however, limited to tables of figures, which at the best of times can be confusing, and at the worst can be downright misleading. This chapter is, therefore, concerned with using the graphical capabilities of the Excel application to draw out and to edit diagrams which represent the results of the model. Eventually (Chapter 10) we shall return to graphical displays as a medium for testing the models, but initially we content ourselves with determining the difference between such specific graphs and the more general simulation results which are being produced by the hydrological model. This chapter explains how to highlight data within an Excel file, and how to choose and edit some of the graphs which are available.

OPENING THE CHART OPTIONS

Graphs within Excel are called 'charts' and data to be graphed must firstly be identified and then transferred to the 'chart option' before detailed editing can commence. The steps are accomplished by:

1. Open the Excel file which contains the spatial model of channel discharge at the six catchments (Figure 5.10).
2. Highlight cells B19 to G19 to specify the data which are to be plotted, in this case the discharge at the six locations.
3. Open the file menu by clicking the menu bar.
4. Choose the new sub-menu from the main file menu.
5. Click the chart pip (Figure 6.1)
6. Click 'OK'.
7. The data are displayed in one of a number of chart formats, according to the default or prefered options which have previously been set.

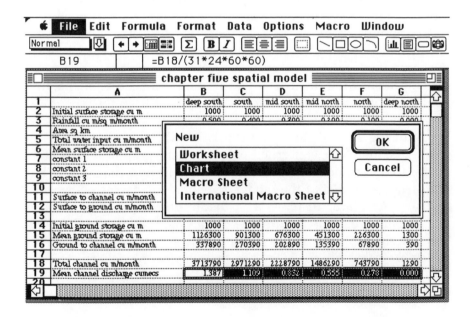

Figure 6.1 *Choosing the chart option from the new and file menus*

DETAILS OF THE CHART OPTION

Excel offers a range of options for displaying the data, and each of these are chosen from the gallery menu on the menu bar. They include:

1. Area charts as shown in Figure 6.2.
2. Bar charts as shown in Figure 6.3 which plot horizontally.
3. Column charts as shown in Figure 6.4 which plot vertically.
4. Line charts as shown in Figure 6.5.
5. Pie charts as shown in Figure 6.6.
6. Scatter charts as shown in Figure 6.7.
7. A range of three-dimensional plots including 3-D area (Figure 6.8), 3-D column (Figure 6.9), 3-D line (Figure 6.10) and a 3-D pie chart.
8. Experiment with each of these by clicking the chosen format from the sub-menu display.

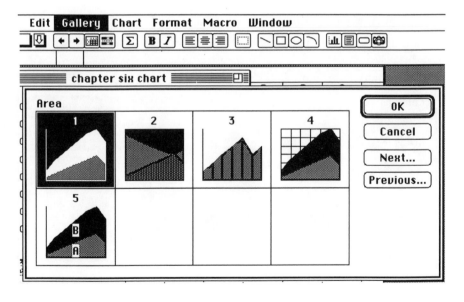

Figure 6.2 *The area option within the gallery menu*

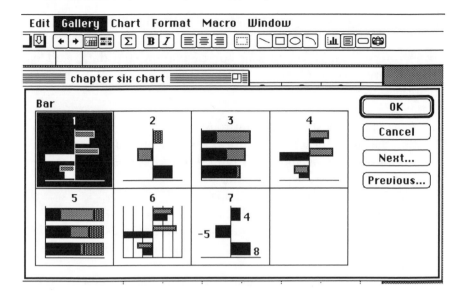

Figure 6.3   *The bar chart options within the gallery menu*

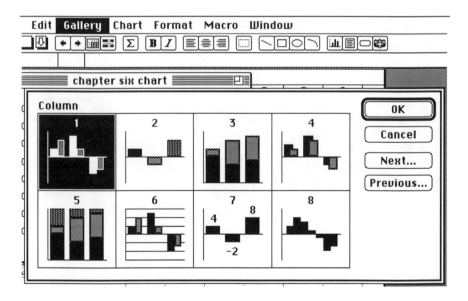

Figure 6.4   *The column chart options*

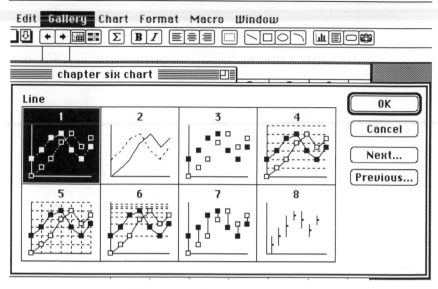

Figure 6.5 *The line chart options*

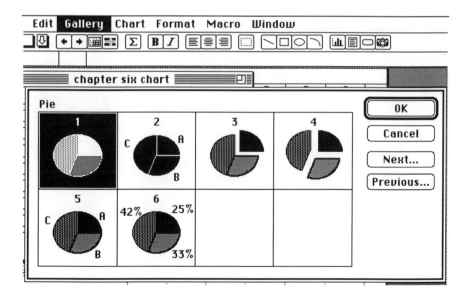

Figure 6.6 *The pie chart options*

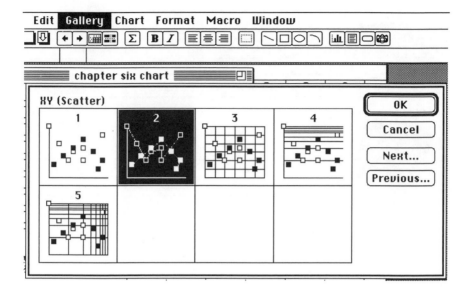

Figure 6.7   *The scatter chart options*

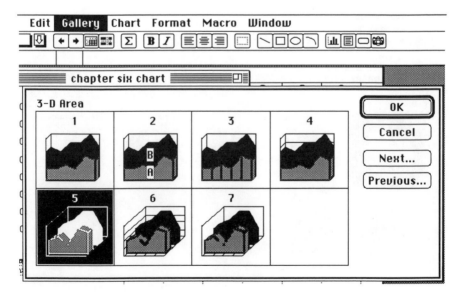

Figure 6.8   *The 3-D area chart options*

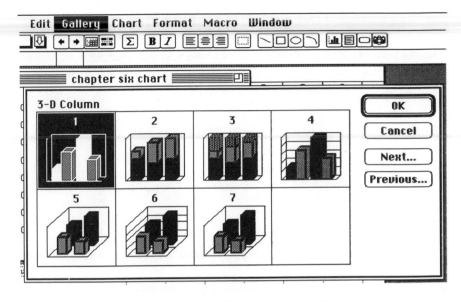

Figure 6.9 *The 3-D column chart options*

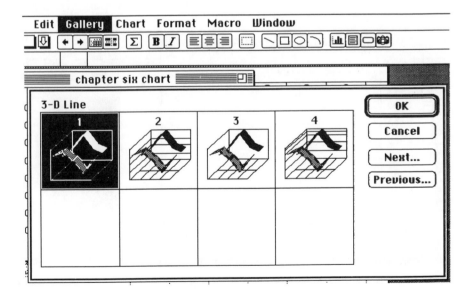

Figure 6.10 *The 3-D line chart options*

LINEAR–LINEAR GRAPHS

The various charting options within Excel were designed for business presentations. Here we are concerned with the proper graphs offered by the line chart option:

1. Keep the chart open for the discharge data.
2. Choose the gallery menu.
3. Choose the Line sub-menu.
4. Choose the linear–linear plot with grid line (option 5) by clicking it on.
5. Click 'OK'.
6. The graph (Figure 6.11) demonstrates the relationship between discharge and distance for the spatial model on a linear–linear display.

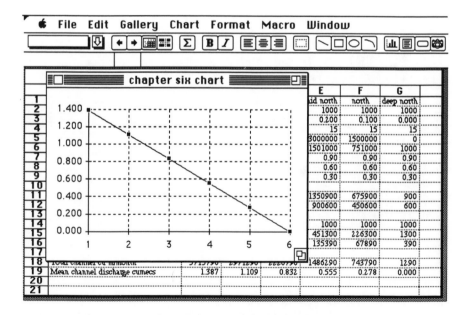

Figure 6.11 *Line chart representation of the model*

EDITING THE CHART

The display in Figure 6.11 can now be edited using other menu options within Excel:

1. Use the size box to stretch the chart over the display screen. Note that Excel correctly resizes both the horizontal and vertical dimensions of the display.
2. Choose the chart menu by clicking the menu bar.
3. Choose the 'attach text' sub-menu in the chart menu by clicking the cursor.
4. Click the chart title pip.
5. Click 'OK'.
6. Type a title (for example 'your name') into the formula bar and then click the tick to enter it onto the screen.
7. Add a category axis (horizontal axis) label in the same way.
8. Add a value axis (vertical axis) label in the same way (Figure 6.12).

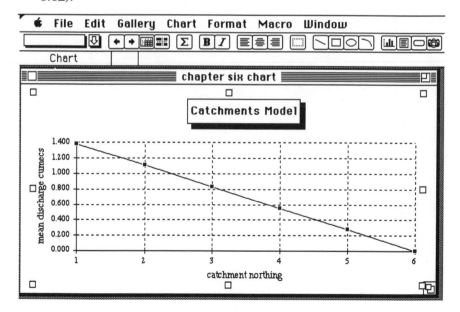

Figure 6.12 *Editing the chart*

9. Note that attributes including the text size, font and orientation can be altered by using the text sub-menu on the Format menu from the menu bar or by double clicking on the appropriate elements of the chart and choosing the font option.

EXAMPLE

1. Follow the steps above to produce a scaled and labelled graph on linear–linear axes of the spatial catchment model. Save the result to disk, and print out a copy.

7
Process-Response Modelling

INTRODUCTION

The previous chapters have been concerned with introducing the Excel spreadsheet. The result was firstly a temporal model with six columns of six formulae which calculated a mean discharge at Catchwater Catchment for the months from January to June for the simplified hydrological system. Secondly a spatial model was developed showing the discharge variations for six catchments having a range of climatic controls. The output from the model was used to draw out and to edit diagrams which represent the results of the model. This chapter is concerned with the construction of a different environmental model. Here we consider the case of boundary layer flow over a desert dune field and see how a model of this phenomenon is erected and coded into Excel.

GENERAL PROCEDURE

The generation and coding of any computerised environmental model involves 15 separate stages. These are:

1 Define the problem
2 Decide if the problem is spatial or temporal

3   Write down the input parameter (space or time)
4   Write down the input constants
5   Write down the processes (formulae)
6   Write down the processes as Excel formulae on paper
7   Write down the output parameters
8   Plan the screen display
9   Format the column width
10   Format the number type
11   Format the number justification
12   Input the text
13   Input the constants
14   Input the formulae
15   Check the model with a calculator

We shall now follow these stages to code the computer model which examines one aspect of the environmental system in more detail. We shall develop a model of boundary layer flow in the wind over a desert dune field.

Define the Problem

The problem in this case is to determine the variation of wind speed with height above the sand surface as illustrated in Figures 7.1 and 7.2.

Decide if the Problem is Spatial or Temporal

Clearly, this problem depends upon height above the bed which is a length scale and therefore this model is spatially dependent.

Write Down the Input Parameters, Space or Time

The input parameter here is the vertical height which is signified by z and will have units of centimetres.

Figure 7.1 *Photograph of anemometers being used to measure wind speed at various heights above the dune surface in the north west Algerian Sahara (JH)*

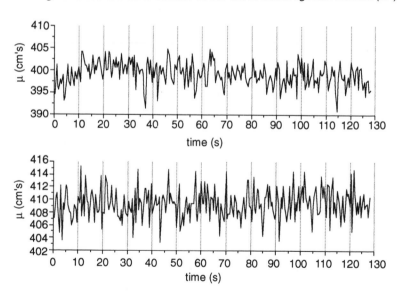

Figure 7.2 *Typical datasets for the wind speed at various heights above the dune surface obtained with the anemometer array shown in Figure 7.1*

Write Down the Input Constants

The input constants here are the wind speed at a height of 100 cm above the surface which is signified by U and will have units of centimetres per second (a second constant will be introduced at a later stage in this example).

Write Down the Processes (Formulae)

Two models are being constructed. In the first (*Model I*), the hypothesis is that the flow speed is directly proportional to the height above the bed, z, and to the upper speed, U. Thus the formula for the flow speed, u is:

$$u = c.U.z$$

where c is the second input constant, and is called the coefficient of proportionality.

In the second model (*Model II*), a similar hypothesis is proposed, but in this example the flow speed is directly proportional to the logarithm of the height so that:

$$u = cU \log(z)$$

Write These as Excel Formulae on Paper

These become:

Model I $\qquad = c * U * z$

Model II $\qquad = c * U * \text{LOG}(z)$

although, of course, the symbols will be replaced by cell references on the spreadsheet.

Write Down the Output Parameter

The output parameter is the local flow speed, u with units of centimetres per second.

Plan the Screen Display

This is done on a blank Excel sheet as shown in Appendix I of this book.

Format the Column Width

Starting from the model.blank file which we used earlier:

1. Drag the display and resize using the size box until the worksheet occupies the whole screen.
2. Highlight all of column A and set the column width to 16 units from the format menu from the menu bar.
3. Highlight all of columns B to K and set the column width to 4 units using the format menu from the menu bar.

Format the Number Type

1. Highlight columns B to K.
2. Set the number type to 0.0 by typing it into the option box within the number sub-menu on the format menu from the menu bar.

Format the Number Justification

1. Highlight columns B to K.
2. Set the justification to right within the alignment sub-menu on the format menu from the menu bar.

Input the Text

1. Type your name into cell A1.
2. Type in the rest of the text and the height values in row 3 as shown in Figure 7.3.

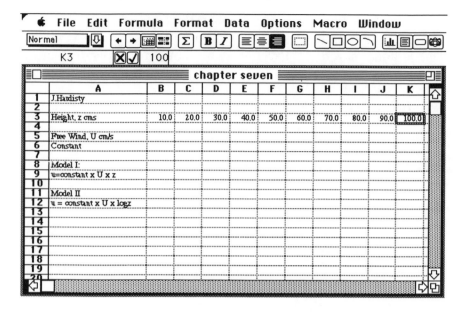

Figure 7.3 *Laying out the spreadsheet for the boundary layer model*

Input the Constants

1. Type the initial free wind speed, 25, into cell B5 and fill right from the edit menu to duplicate into columns C to K.
2. Type the initial coefficient of proportionality, 0.1, into B6 and fill right.

Input the Formulae

1. For Model I, type '=B5*B6*B3' into cell B9.
2. Highlight cell B9 and fill right to column K.
3. Check that Excel has correctly re-referenced the formula in each cell. For example cell K9 should now contain '=K5*K6*K3' and so on (Figure 7.4).
4. For Model II, type '=B5*B6*LOG(B3)' into cell B12.
5. Highlight cell B12 and fill right to column K.
6. Again check that Excel has correctly re-referenced the formula in each cell. In this case cell K12 should contain '=K5*K6*LOG(K3)' and so on (Figure 7.5).

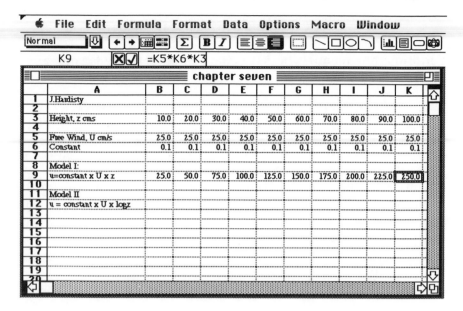

Figure 7.4 *Entering the formulae for Model I onto the spreadsheet*

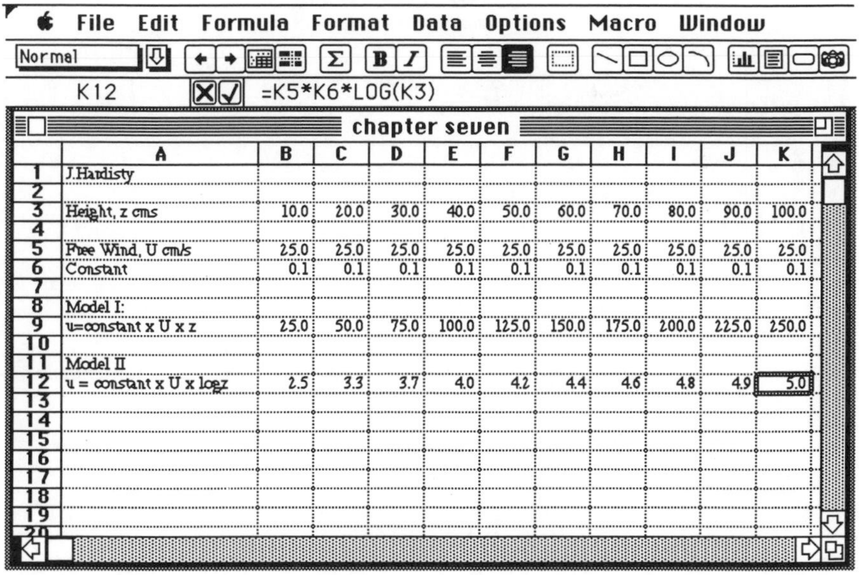

Figure 7.5 *Entering the formulae for Model II onto the spreadsheet*

Check the Model with a Calculator

Perform a few manual checks on the operation of the formulae.

RUNNING THE MODEL AND GRAPHING THE RESULTS

Now use the chart option from the new menu to draw a graph of the variation of flow speed according to Model II with the second option in the line graph sub-menu from the gallery menu on the menu bar. Examine 'what if' situations by varying the surface flow speeds in row 5, but remember to duplicate the new flow speeds into all columns (Figure 7.6).

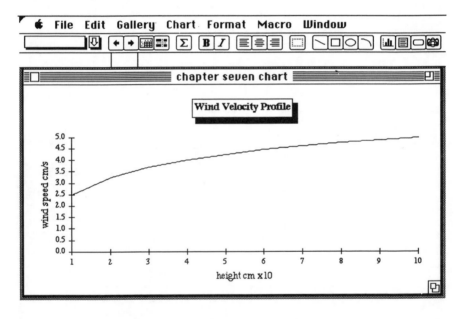

Figure 7.6   *Line chart of the final model*

EXAMPLE

1.   Follow the steps above to complete the models.
2.   Set the free wind speed to 35 cm s^{-1} in all columns.

3. Vary the coefficient of proportionality until the flow speed at a height of 100 cm above the bed in Model II is the same as the free wind speed.
4. Plot a graph of the results and print out the graph.
5. Print out your final worksheet of the model in this form.

8
Stochastic Modelling

INTRODUCTION

The previous chapters have been concerned with introducing the Excel spreadsheet and with coding temporal and spatial hydrological models. Chapter 7 considered the first of the types of analysis which were introduced in the first chapter and developed a simple, deterministic, process-response model for boundary layer flow. It did not include any feedbacks since the boundary layer flow (the output from the model) was not used to influence the free stream velocity (the input). It was deterministic in that the output was a unique function of the input. This chapter is concerned with the construction of a different geographical model. Here we consider the case of the development through time of local temperatures and see how a model of this system is erected and coded into Excel. We shall follow the 15 stages to code the computer model of temperature changes. In particular, the models covered in the preceding chapters have all been fully deterministic, that is to say a given input always produces the same output. In this chapter we introduce the notion that such is not always the case; many natural processes are random and prediction therefore requires a stochastic element. This stochastic element can be thought of as a random effect on the process or transfer function (Chapter 1) so that slightly different outputs are produced on each occasion. We shall introduce the random element through one of the in-built functions in Excel and the chapter therefore begins with a more general introduction to the Excel functions.

EXCEL FUNCTIONS

One of the most powerful features of the Excel package is the large number of in-built functions which are available for use in the formulae. We have already met the simple arithmetic functions like +, -, * and / and, in the last chapter we used the base 10 logarithm function, LOG. The Excel package contains almost 500 such functions and a number of these are listed in Appendix II. A number of Excel functions are used in the present chapter and they are introduced in the present section. In general, the functions can be divided into nine groups which are given here and each is followed by a number of examples:

1. Mathematical functions. Perform mathematical operations such as addition, subtraction, exponentiation and so on.

2. Statistical functions. Calculate statistical properties of groups of numbers such as the mean, maximum, standard deviations and so on.

3. Database functions. In addition to being a spreadsheet, the Excel application has powerful database capabilities in which data can be stored and retrieved. These operations are controlled by the database functions.

4. Trigonometric functions. Perform standard trigonometric transformations on the circular and hyperbolic functions such as sine, cosine, hyperbolic sine and so on.

5. Logical functions. Based upon the Boolean algebra, there are a wide range of IF, AND and NOR functions available.

6. Text functions. In addition to working with numbers, text characters and text strings can be manipulated using these functions.

7. Financial functions. Reflecting Excel's prime usage in the business community, these functions provide, for example, interest rate calculations.

8. Date functions. Again for business usage, Excel spreadsheets are frequently used on a day-to-day basis and need to determine, for example, the number of days which have elapsed in a given period. The application provides for such calculations through the various date functions which are available. The database, financial and date functions are not covered in this book.

9. Special functions. This is a catch-all compartment for operations which do not fit under the other classifications.

A number of the functions in each of these categories are introduced in the following sections.

Mathematical Functions

The mathematical functions in the following examples operate on 'number' which can be a cell reference, a constant or the result of any other functions or calculations. Seven mathematical functions are listed here and further examples are given in Appendix II.

- ABS(*number*): returns the absolute, i.e. the positive, value of number. Thus if *number* equals 15 or -15, the function returns (+) 15.
- EXP(*number*): returns the exponential of *number*, i.e. e^{number}. Since the value of e is approximately 2.72, then if *number* equals 2 then the result is 7.39 or if *number* is A1 and cell A1 contains -3 then the result is 0.05

- INT(*number*): returns the integer value of *number*. Thus if *number* 1.5678 then the result is 1.
- LN(A1): returns the natural (i.e. base e) logarithm of the parameter. Thus if cell A1 contains 6 then the result is 1.7918.
- LOG(A1): returns the logarithm (i.e. base 10) of the parameter. Thus if cell A1 contains 7.4 then the result is 0.8692.
- PI(): returns the value of π. Note that the empty brackets must be included in this function.
- RAND(): returns a random number between and including 0 and 1. Note that every time the model is recalculated a different value is returned. Thus four values for RAND() might be 0.743, 0.615, 0.126 and 0.520.

Statistical Functions

The statistical functions provide results from a range of numbers defined in the parameter statement. The parameter statement can be in one of two forms. Either the cells to be considered can all be listed and separated by commas or, if they all fall within one column or row, then the two extreme cells are listed and separated by a colon. Thus if the function is to operate on the contents of, for example, the five cells A1, B1, C1, D1 and E1 then the cells can be specified by either (A1, B1, C1, D1, E1) or by (A1:E1). The following examples will simply use (A1, B1......) to indicate that the user must specify using one or other of the above methods. Five statistical functions are introduced here, and further functions are given in Appendix II.

- AVERAGE (A1, B1...): returns the mean value of the conents of the specified cells.
- MAX(A1, B1...): returns the maximum value of the contents of the specified cells.
- MIN(A1, B1...): returns the minimum value of the contents of the specified cells.
- STDEV(A1, B1...): returns the standard deviation of the contents of the specified cells.
- SUM(A1, B1...): returns the arithmetic sum of the contents of the specified cells.

Trigonometric Functions

All of the trigonometric functions operate on radian measures and, therefore, if the parameter is in degrees it should be multiplied by PI()/180 (because there are π radians in 180°) in order to convert to radians. In the following, *'number'* can be a cell reference, a constant or the result of any other functions or calculations. Six trigonometric functions are listed here and further examples are given in Appendix II.

- COS(*number*): returns the cosine of *number*.
- SIN(*number*): returns the sine of *number*.
- TAN(*number*): returns the tangent of *number*. Thus:

$$=TAN(PI()*A1/180)$$

returns the tangent of A1, where A1 is in degrees.
- ACOS(*number*): returns the angle the cosine of which is *number*.
- ASIN(*number*): returns the angle the sine of which is *number*.
- ATAN(*number*): returns the angle the tangent of which is *number*.

Logical Functions

There are a range of logical functions available within Excel, of which the most commonly used is the conditional IF statement.

- IF(logical statement, value if true, value if false): returns the value if true or value if false corresponding to the results of the logical statement. For example, suppose a spreadsheet contained examination results in cell A17 and you wanted the word 'PASS' to appear in cell B17 if the mark was greater than 56 but the word 'FAIL' to appear if the mark were equal to or less than 56, then the following conditional statement could be entered into cell B17:

$$= IF (A17>56,"PASS","FAIL")$$

Note that the text is contained in quotation marks and that commas must be used to separate the various elements of the function.

Text Functions

It should by now be clear that Excel treats the contents of cells as one of only three types of information: numbers, text or formulae. The text functions are used to operate on streams of characters called strings in the same way as the mathematical and other functions operate on numbers. In the following examples *string$* will be used to indicate a piece of text, or a cell reference which contains a piece of text, or a piece of text which results from other functions. Four text functions are listed here and further examples are given in Appendix II.

- LEN (*string$*): returns a result equal to the number of characters in *string$*. For example, if string$ is the cell reference A1 and cell A1 contains the word 'INSTRUCTIONS' then LEN(A1) returns 12.
- MID(*string$, start character, number of characters*): returns a piece of text having *number of characters* and starting at (and including) *start character* in *string$*. For example, if string$ is the cell reference A1 and cell A1 contains the word 'INSTRUCTIONS' then MID(A1,9,3) returns the word 'ION'.

It is important to remember that Excel must differentiate between numbers and text and must therefore be told when to convert the one to the other. Two further text functions are used to undertake these conversions:

- TEXT(*number*): returns a piece of text or word equivalent to *number*. For example, you may wish to know how many digits are contained in the year of the Battle of Hastings. This could be accomplished by:

$$= LEN(TEXT(1066))$$

Simply writing '=LEN(1066)' would produce an error because 1066 is a number and as such cannot be operated upon by a text function.
- VALUE(*string$*): this is the opposite of the TEXT function and returns a number equivalent to the word or string of text *string$*.

Special Functions

There are two particularly useful functions which fall into this category and are based upon the cell reference where *cell reference* can be, for example, A1 or A2 or the result of another function:

- COLUMN(*cell reference*): returns the column letter of the cell reference.
- ROW(*cell reference*): returns the row number of cell reference. For example

$$=ROW(IF(A1>A2,A1,A2)$$

returns '1' if the contents of A1 are greater than those of A2 and otherwise returns '2'.

TEMPERATURE MODELLING

The generation and coding of any computerised environmental model involves 15 separate stages detailed in the preceding chapter. We shall now proceed to follow these steps to erect a stochastic model for temperature variations at Edinburgh in Scotland.

1. *Define the problem.* The problem in the present case is to construct a model for minimum daily temperatures at Edinburgh for December 1990.
2. *Decide if the problem is spatial or temporal.* The problem is, quite clearly, temporal and deals with daily temperature variations.
3. *Write down the input parameter (space or time).* We shall assume that, for the purposes of the present model, the input parameter will be the number of hours of sunlight on each day.
4. *Write down the input constants.* There will be some constant which represents the 'thermal capacity' of Edinburgh and another which represents the 'absolute minimum' for the site.
5. *Write down the processes (formulae).* We shall assume that the minimum temperature is equal to the absolute minimum plus a stochastic element related to the thermal capacity and the number of hours of sunlight, i.e.:

Minimum Temperature = Absolute Minimum +
$$R \times \text{Thermal Capacity} \times \text{Hours of Sunlight}$$

or in symbols:

$$T_{min} = T_{abs} + R \times C \times S$$

where T_{min} is the minimum temperature in °C,

T_{abs} is the absolute minimum temperature,

R is a random element which accounts for meteorological factors,

C is the constant representing the thermal capacity, and

S is the number of hours of sunlight.

6. *Write down the processes as Excel formulae on paper.* The Excel equivalent of the formula is:

$$= \text{CELL(A1)} + \text{RAND()*CELL(A2)*CELL(A3)}$$

where cells A1, A2 and A3 will contain the appropriate values of the absolute minimum temperature (which is a constant), the thermal capacity (which is the second constant) and the hours of sunlight. The RAND() function was introduced earlier.

7. *Write down the output parameters.* The output parameter will be the minimum daily temperatures for the month.

8. *Plan the screen display.* The screen display will have a control panel at the top and will continue with the convention that temporal variation runs horizontally across the screen. The date is entered in row 4 from column B to column AE as shown in Figure 8.1.

9. *Format the column width.* It is clearly not possible to display all 31 days on the screen, but progress is achieved by setting the text to a relatively small font and size and then reducing the column widths as much as is practicable. In the present example:

   (i) *Set display font:* Click on the small box to the left of the column headers so that the whole of the spreadsheet is highlighted. Choose a font from the format menu and choose 10 point Times.

12. *Input the text.* The text is used to explain the layout of the model. Type the following words into the appropriate cells:

A4	Date	A5	Sun Hrs
A6	Random	A7	Min Temp
C2	Absolute Minimum	H2	Thermal Constant

You will note that Absolute Minimum and Thermal Constant extend beyond the right-hand margin of their cells. This is another advantage of Excel and overcomes the problem of narrow cell widths in the present model. The result is shown in Figure 8.2.

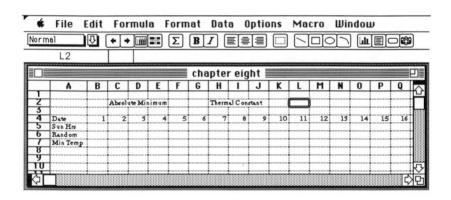

Figure 8.2   *Inputting text into the model*

13. *Input the constants.* There are two constants and then the appropriate sunshine figures to be input. Taking initial values, type 5 into cell F2 for the absolute minimum temperature and 1.1 into cell K2 for the thermal constant. Then input the hours of sunlight for Edinburgh for June 1990 from Table 8.1 (Figure 8.3).

14. *Input the formulae.* There are only two sets of formulae in the model, and the first simply illustrates the use of the random function:

Table 8.1 *Hours of sunlight for Edinburgh, June 1990*

Date	Cell	Value	Date	Cell	Value
1	B5	1.4	16	Q5	12.5
2	C5	9.5	17	R5	6.6
3	D5	7.3	18	S5	1.6
4	E5	6.4	19	T5	8.2
5	F5	0.1	20	U5	3.8
6	G5	0.2	21	V5	1.9
7	H5	10.6	22	W5	0.2
8	I5	5.7	23	X5	11.9
9	J5	6.6	24	Y5	2.3
10	K5	4.4	25	Z5	11.0
11	L5	3.0	26	AA5	2.4
12	M5	3.4	27	AB5	0.9
13	N5	2.8	28	AC5	11.3
14	O5	0.1	29	AD5	1.0
15	P5	4.6	30	AE5	1.0

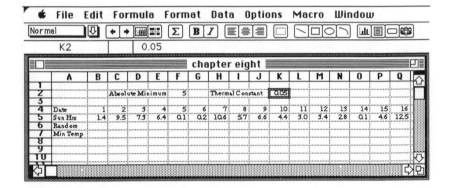

Figure 8.3   *Inputting numbers into the temperature model*

(i) *Random number generator*: type '=RAND()' into cell B6 and fill right to copy the formula into cells B6 to AE6.

(ii) *Calculating the minimum temperature*: type the model formula into cell B7:

$$= \$F\$2 + B6*\$K\$2*B5$$

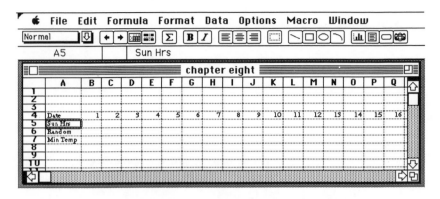

Figure 8.1 *Setting column width to three units in the temperature model*

   (ii) *Set column width*: Highlight columns B to AE and then choose a column width from the format menu and set to 3 units. The result is shown in Figure 8.1.

10. *Format the number type.* Different formats are required for the four rows of numbers:

   (i) *Date*: the dates are presented in integer format. Highlight cells B4 to AE4 and choose integer (0) format from the number sub-menu within the format main menu.

   (ii) *Hours of sunshine*: the hours of sunshine are to be given to one decimal place. Highlight cells B5 to AE5 and choose 0.0 from the number sub-menu by typing '0.0' and clicking 'OK'.

   (iii) *Random number*: the stochastic element in the model will be provided by the RAND() function and we will display the result to two decimal places. Highlight cells B6 to AE6 and choose 0.00 from the number sub-menu.

   (iv) *Minimum temperature*: Finally, the output from the model will be taken to be to the nearest whole number of degrees. Highlight cells B7 to AE7 and choose an integer format from the number sub-menu.

11. *Format the number justification.* We shall apply the same number justification to all cells. Highlight B4 to AE7 and choose 'right' from the alignment sub-menu under the format menu.

and fill right to copy the formula into cells B7 to AE7. There are a number of important aspects of this formula to note. Firstly B5 and B6 contain the hours of sunshine and the random element respectively, these relative references have changed as the formula was filled right. However, F2 and K2 have not changed as the formula was filled right and this alternative form is called absolute referencing. It is designed to be used in examples like the present where a cell reference needs to be fixed because this obviates the need to copy constants through the model. The result is shown in Figure 8.4 and represents the completed model. Again, this is a stochastic model so that the result varies every time the model is run and generates new results for the random function. The model can be re-run by choosing 'calculate now' from the option menu so that no two iterations produce identical results. Save a copy of the model onto your disk for use in the next chapter.

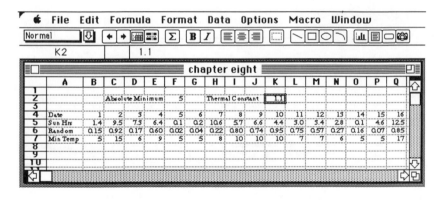

Figure 8.4 *The completed temperature model*

15. *Check the model with a calculator.* Without a random number generator on the calculator it is, of course, impossible to check the model; however, use the values displayed on one of the

days to check that the answer is as required. This can usefully be accomplished by introducing another Excel feature. Choose the display sub-menu from the options menu as shown in Figure 8.5. Click on the formula box and 'OK'. The spreadsheet is then displayed with the formulae as opposed to the results of the formulae visible on the screen as shown in Figure 8.6.

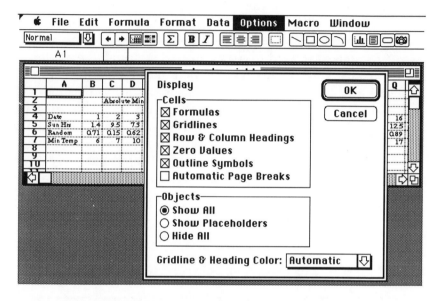

Figure 8.5 *Choosing to display formulae on the spreadsheet*

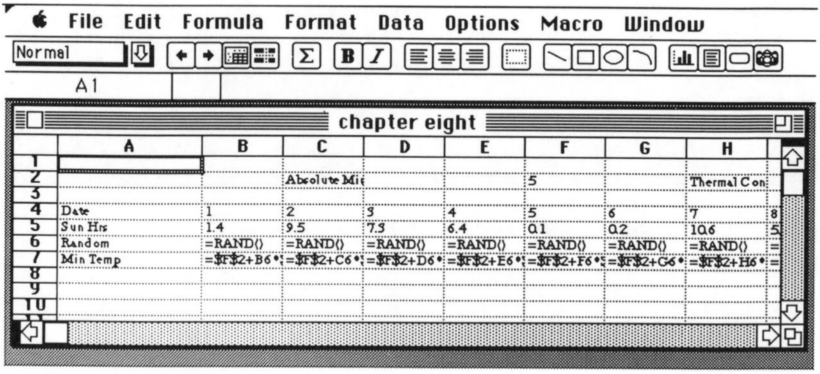

Figure 8.6 *The spreadsheet with formulae as opposed to numbers displayed*

EXAMPLE

Input the Edinburgh temperature model as described above and plot a graph of minimum temperature variation through the month (Figure 8.7).

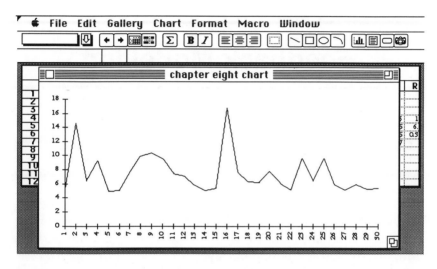

Figure 8.7 *One example of the minimum temperatures predicted by the stochastic model*

9
Feedback Modelling

INTRODUCTION

The previous chapter was concerned with the construction of a stochastic model and considered the case of temperature variations through time. Earlier, boundary layer flow over a dune surface was modelled as an example of the genre of models known as process-response models (see Chapter 1). More realistically the outputs of an environmental system or subsystem affect the inputs and processes within the system, and this type of approach is known as feedback modelling. In this chapter a feedback model is introduced which represents the evolution of a hillslope profile through time and we shall see how a model of this system is erected and coded into Excel. In particular, feedbacks can be positive, in which case the changing input is amplified by the system leading to instabilities, or negative, in which case change tends to be attenuated and the system stabilises. We shall see that, with a simple change to the process formula, the hillslope model can simulate either positive or negative feedbacks so that irregularities in the slope profile are either accentuated or are smoothed through time.

HILLSLOPE MODEL

Following the procedure which was outlined in Chapter 7 and was utilised in Chapter 8, the construction of the model can be divided into 15 stages:

1. *Define the problem.* The problem in this case is to determine the evolution of a hillslope profile.
2. *Decide if spatial or temporal.* Clearly, this problem depends upon the evolution of the system which is a time scale and therefore this model is temporal or time dependent.
3. *Write down the input parameters, space or time.* The input parameter here is the passage of time which is signified by t and will have units of years.
4. *Write down the input constants.* The input constant here is the coefficient of proportionality in the proposed model which is called the response constant, C. Furthermore the hillslope is to be separated into 15 sites, the elevation of the first of which is to be fixed at 1000 m.
5. *Write down the processes (formulae).* In this model, the hypothesis is that the erosion at each site is directly proportional to the difference in height between that site and the next one further up the slope (i.e. it is proportional to the gradient of the hillslope). Using *t* for time and *n* for site (where $n = 1$ is the top site) the formula is:

Height at n = Previous height at
$$n - C(\text{Difference in height from site } n - 1 \text{ to } n)$$

or in symbols:

$$Z_1 = Z_2 - C(Z_3 - Z_4)$$

where C is called the coefficient of proportionality, Z_1 is the new height at a particular site, Z_4 was the previous height at that site and Z_3 was the previous height at the previous site.

6. *Write this as an Excel formula on paper.* This becomes:

$$= Z_2 - C * (Z_3 - Z_4)$$

although, of course, the symbols will be replaced by cell references on the spreadsheet.

7. *Write down the output parameter.* The output parameter is the new height, Z_1 with units of metres.
8. *Plan the screen display.* This is done on a blank Excel sheet as shown in Appendix I of this book.
9. *Format the column width.* Starting from the model.blank file which we used earlier:

   (i) Move the display and resize using the size box until the worksheet occupies the whole screen.
   (ii) Highlight all of column A and set the column width to 16 units from the format menu on the menu bar.
   (iii) Highlight all of columns B to K and set the column width to 5 units using the format menu from the menu bar.

10. *Format the number type.*

   (i) Highlight columns B to K.
   (ii) Set the number type to 0 by choosing this option within the number sub-menu on the format menu from the menu bar.

11. *Format the number justification.*

   (i) Highlight columns B to K.
   (ii) Set the justification to right within the alignment sub-menu on the format menu from the menu bar.

12. *Input the text.*

   (i) Type your name into cell A1.
   (ii) Type in the rest of the text and the height values in column B as shown in Figure 9.1

13. *Input the constants.*

   (i) Type the response constant, 0.1, into cell A3. We shall be using absolute addresses in the formulae and therefore will not need to duplicate this constant in the worksheet; simply refer to *A3*.

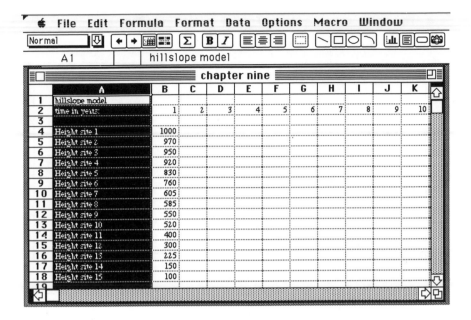

Figure 9.1 *Laying out the hillslope model and inputting text*

(ii) It is useful at this point to plot the initial hill profile by highlighting column B from rows 3 to 18 and choosing the chart option, and the fourth line graph option from the gallery menu. The result is shown in Figure 9.2.

(iii) Do not close the chart, simply click on the spreadsheet to return it to the front of the display.

(iv) Highlight cell B4 and row 4 from column B to K. Fill right from the edit menu to place the fixed upslope, 1000 m, height into each of these cells.

14. *Input the formulae.*

(i) Type '=B5-A3*(B4-B5)' into cell C5 as shown in Figure 9.3.

(ii) Highlight cell C5 and across the row to K5 then fill right from the edit menu.

(iii) Check that Excel has correctly re-referenced the formula in each cell. For example, cell K5 should now contain '=J5 - A3*(J4-J5)'.

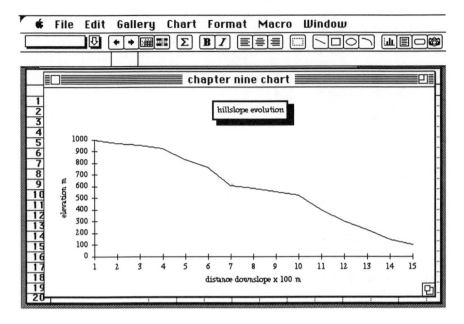

Figure 9.2   *Chart of the initial hillslope*

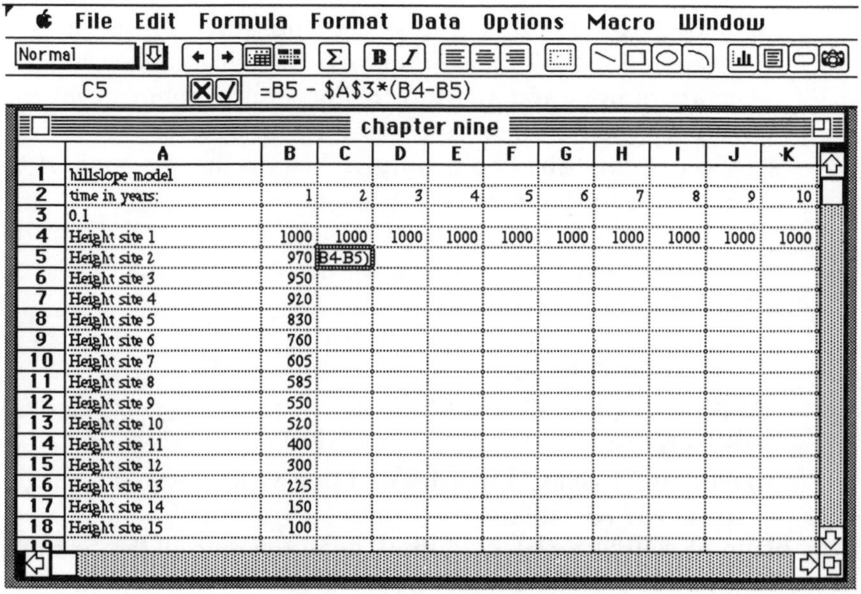

Figure 9.3   *Inputting formulae*

(iv) Highlight cell C5 and fill down to C18. Repeat for columns D to K.

(v) Again check that Excel has correctly re-referenced the formula in each cell. In this case cell K18 should contain '= J18 - A3*(J17-J18)'.

15. *Check the model with a calculator.* The model is now completed as shown in Figure 9.4. Check some of the calculations with a calculator.

🍎 File Edit Formula Format Data Options Macro Window										
Normal ⬇ ← → 📇 ⊞ Σ **B** *I* ▤▤▤ ⬚ ◥□○◣ ⽥▤○❀										
K18		=J18-A3*(J17-J18)								

	A	B	C	D	E	F	G	H	I	J	K
1	hillslope model										
2	time in years:	1	2	3	4	5	6	7	8	9	10
3	0.1										
4	Height site 1	1000	1000	1000	1000	1000	1000	1000	1000	1000	1000
5	Height site 2	970	967	964	960	956	952	947	942	936	929
6	Height site 3	950	948	946	944	943	941	940	940	940	940
7	Height site 4	920	917	914	911	907	904	900	896	892	887
8	Height site 5	830	821	811	801	790	778	766	753	738	723
9	Height site 6	760	753	746	740	734	728	723	718	715	713
10	Height site 7	605	590	573	556	537	518	497	474	450	423
11	Height site 8	585	583	582	583	586	591	598	608	622	639
12	Height site 9	550	547	543	539	534	529	523	516	506	495
13	Height site 10	520	517	514	511	508	506	503	501	500	499
14	Height site 11	400	388	375	361	346	330	312	293	272	250
15	Height site 12	300	290	280	271	262	253	246	239	233	229
16	Height site 13	225	218	210	203	197	190	184	177	171	165
17	Height site 14	150	143	135	127	120	112	104	97	88	80
18	Height site 15	100	95	90	86	82	78	74	71	69	67
19											

Figure 9.4 *The completed hillslope model*

RUNNING THE MODEL AND GRAPHING THE RESULTS

Now begin at cell B4 and highlight all columns and all rows to cell K18. Choose copy from the chart menu and then reactivate the chart by clicking the chart name on the windows sub-menu on the main menu bar. Once the chart is active, click the paste option on the edit menu and all of the profiles are displayed simultaneously as shown in Figure 9.5. This is an example of positive feedbacks with all of the irregularities being amplified during erosion.

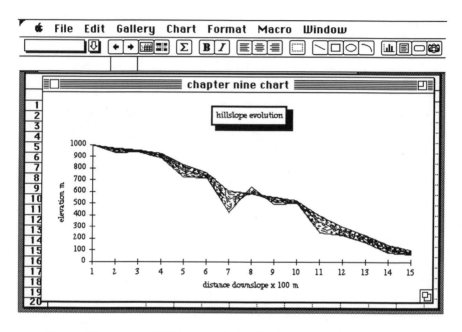

Figure 9.5 *Graph of hillslope evolution for the positive feedback model*

EXAMPLE

1. Follow the steps above to complete the model.
2. Change the model so that erosion is proportional to the difference in heights between the present and the next site, rather than to the difference in height between the present and previous sites.
3. Plot the results (as shown in Figure 9.6) and assure yourself that you have produced a negative feedback model, with erosion tending to smooth out the irregularities in the hillslope profile.

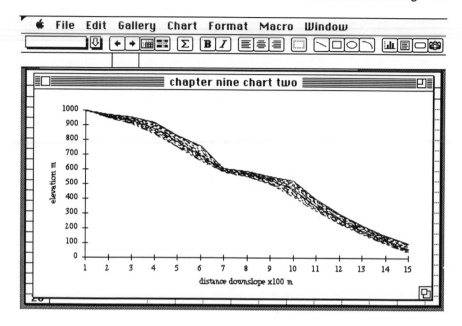

Figure 9.6 *Graph of hillslope evolution for the negative feedback model*

10
Validation

INTRODUCTION

The introductory chapters in Part I of this book emphasised some of the philosophical background to computerised environmental modelling. In particular they explained that modelling is a tool with which a better understanding of environmental systems can be constructed. Later chapters have demonstrated how a variety of models may be constructed and compared simple process-response models with stochastic and feedback approaches. Here we return to wonder whether the models offer an improved understanding in the sense which was originally intended. This is best achieved by expanding upon the terms and concepts of the modelling process which have been introduced throughout the book.

Chapter 1 explained that the systems approach involved four phases (Huggett, 1980):

1 *The lexical phase;* involving the definition of the boundaries, the definition of the state variables and the choice of values for the variables. This aspect is covered in the definition of the problem and of the variables which was introduced in Chapter 7 as the

first four of 15 practical steps towards model construction.

2   *The parsing phase;* involving the definition of the relationships between the state variables. This is covered by the fifth of the 15 steps introduced in Chapter 7.

3   *The modelling phase;* involving firstly the actual model construction and secondly operationalising or running the model. This is covered by the remainder of the steps introduced in Chapter 7 and, in the present instance, has involved the use of the spreadsheet application to achieve numerical solutions to the formulae.

4   *The analysis phase;* involving the comparison of the model's prediction with the observed (e.g. field) data. This phase can be the most demanding and time consuming of the whole process and can lead to some very sophisticated work with advanced coding and data analyses. Although the details are clearly beyond the scope of the present book, a worthwhile overview can be gained by considering the analysis of the stochastic temperature model which was developed in Chapter 8.

In practice, the analysis of a numerical model can itself be divided into three phases which will usually overlap:

4(a)   *Validation phase;* involving the comparison of the model's prediction with the observed (e.g. field) data. This can vary from a simple ratiometric comparison through more or less sophisticated correlation analyses to determine the goodness of fit (or otherwise) and finally may incorporate considerable model modification as described in the other two analysis phases below. The objective is to determine how well the model outputs compare with observation. In the present chapter we shall compare the time series of temperatures at Edinburgh which were produced in Chapter 8 with data obtained by the Meteorological Office.

4(b)   *Sensitivity analysis phase;* involving a close inspection of the effect of changes in either the input parameters or the process formulae on the output of the model. The objective is to concentrate attention on the most important or critical aspects of the system or subsystem.

4(c)   *Optimisation phase*; involving changes to 'constants' within the model so that the output more closely resembles the field data at the expense of explanation. Quite frequently optimisation of the model can provide an experimental test bed to determine immeasurable or unmeasured effects and processes.

Each of these three phases is considered separately within the present chapter although it should be emphasised that each is but one part of the final, analysis phase in Huggett's classification. It must be stressed that the present chapter provides a simple introduction to what is an important (if not the most important) part of computerised environmental modelling and further details can be found in any of the more advanced texts (e.g. Medley, 1982; Woldenberg, 1985; Ahnert, 1987; Anderson, 1988; Lakhan & Trenhaile, 1988; Farmer & Rycroft, 1991).

VALIDATION

The processes of validation are those by which we find out how well the model compares with the real world prototype and begin to identify changes which may improve its operation. Take, for example, the stochastic model of temperature variations that was developed in Chapter 8. The actual observed temperature variations for June 1991 at Edinburgh are given in Table 10.1. Clearly, the whole process depends critically upon the quality of the real world data which are being used and data quality and appropriateness is an important aspect of model analysis. The reader is referred to, for example, Thornes & Brunsden (1977) for a discussion of the practical problems associated with temporal measurement in one aspect of the environmental sciences and numerous other texts are available for particular disciplines. The process of validation means setting the observations against the predictions of the model and the process of optimisation means changing the parameters within the model so that its predictions come closer to the observations. This is, in fact, a very necessary and very sophisticated aspect of all environmental modelling work; here we shall keep things relatively simple in order to illustrate the procedures.

Table 10.1 *Sunshine and minimum temperatures Edinburgh, June 1991*

Date	Temperature (°C)	Sun (h)	Date	Temperature (°C)	Sun (h)
1	5	14.3	16	8	0.4
2	7	0.4	17	7	2.8
3	1	7.4	18	7	1.7
4	2	5.4	19	7	10
5	0	12.0	20	4	5.2
6	3	1.9	21	6	0.1
7	7	11.1	22	8	4.3
8	3	3.1	23	8	2.5
9	9	5.0	24	7	3.9
10	10	7.8	25	11	2.5
11	6	2.3	26	10	8.7
12	9	3.8	27	9	1.3
13	7	0.4	28	9	8.2
14	8	7.6	29	9	6.7
15	7	1.6	30	12	4.2

The input representing hours of sunlight (which was not included in the original model) has been entered into the temperature model in accordance with Table 10.1 as shown in Figure 10.1. Validation proceeds on the basis of the 'last graph' concept in which a graph is plotted of the model's predictions on the horizontal or *x*-axis and of the observations on the vertical or *y*-axis as shown in the diagram. It is clear that, although a general trend exists (i.e. the predicted temperatures are generally higher on days having higher observed temperatures), the relationship is not particularly encouraging. A perfect fit would have all points falling on a straight line through the origin and with a unity gradient. The result is hardly suprising because the model was *very* simple and because a stochastic element was included. The modeller should not, however, become disillusioned. The model may be perfectly adequate for the chosen purposes in which case it can be accepted and operationalised to address the problem in hand. Alternatively, it can be improved by either sensitivity analysis or optimisation procedures as described below.

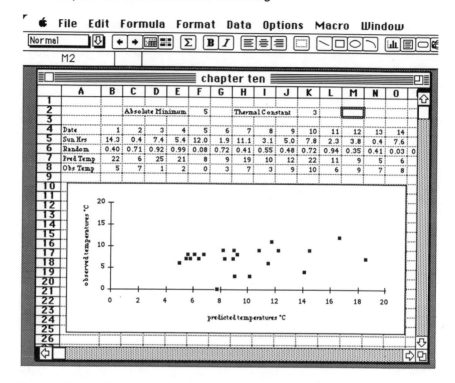

Figure 10.1 Validation of the minimum temperature model by comparing the predicted temperatures with observed data

SENSITIVITY ANALYSIS

Sensitivity analysis means, quite literally, determining how sensitive the output from the model is to either the inputs (called *parameter sensitivity*) or to the formulae or transfer functions (called *process sensitivity*). For example, the beach model that was described in Chapter 1 predicted a beach response which was highly dependent upon incoming waves but almost unaffected by sediment grain size. The model was sensitive to the input parameter (the waves). Each type of sensitivity analysis will now be explained using the minimum temperature model.

Parameter Sensitivity

The use of the spreadsheet for environmental modelling quickly enables simple parameter sensitivity analyses. For example, we could change the model to report the mean and variance of the predicted values as indicators of output. Type 'Mean Temperature' and 'Temperature Variance' into cells C3 and H3 respectively, and then, using the additional functions introduced in an earlier chapter, the following formulae:

$$= \text{AVERAGE (B7:AE)} \qquad \text{into cell F3}$$
$$= \text{STDEV (B7:AE)} \qquad \text{into cell M3}$$

Note that the colon can be used to define an array of cells between two end members rather than listing all of the cells with comma separators. Now, change the inputs, for example increase the absolute minimum in cell F2 by 20% and observe the change in the output. In the present example of a stochastic model it will be necessary to run the model a number of times with each of the different inputs to account for the random element. Other parameters can be analysed in a similar way to determine, in percentage terms, the sensitivity of the model to given changes in each one.

Process Sensitivity

Firstly, the processes which are implicit in the model could be examined using a similar analysis. In the present case there appears to be lower observed temperatures than predicted, so that an improved formula might be incorporated into the model. There is a linear dependence of the predicted temperature on the thermal capacity and on the hours of sunlight. Thus the formula could be gradually adjusted to reduce the predicted temperatures and the resulting output changes observed. This is a trivial problem with the minimum temperature model which, essentially, contains only one formula. However, process sensitivity becomes more involved and more interesting when more than one process is represented.

In either case, experimenting with parameter values or processes provides the operator with a useful insight into the operation of the model and may well lead to a redesign of the whole model. Sensitivity analyses quantify such work.

OPTIMISATION

There are two common types of model optimisation. Firstly the so-called 'constants' in the model can be massaged to improve the relationship between predictions and observations. In the present example a 'thermal constant' of three entered into cell K2 appears to provide the best spread of results. Each of the six models detailed in the following chapters has been parameterised in this fashion.

Alternatively, the trend of the model can be accepted but a final empirical relationship is devised to bring the results closer to the observations. The simplest relationship would be a straight line of the form:

$$\text{Observed Result} = m \text{ (Predicted Result)} + c \qquad (10.1)$$

where the gradient of the line, m, (taken from the model graph) and the intercept, c, are determined by any one of a number of statistical techniques. The validation plot shown in Figure 10.1 is deliberately drawn with the observations on the vertical, y-axis, to enable this process. More sophisticated techniques can be employed, but the simple straight line will generally suffice for the present purposes.

REFERENCES

Ahnert, F. (ed.), 1987. *Geomorphological Models: theoretical and empirical aspects.* Catena Supplement, 10, Catena Verlag, Cremlingen.

Anderson, M.G. (ed.), 1988. *Modelling Geomorphological Systems.* Wiley, Chichester.

Farmer, D.G. & Rycroft, M.J. (eds), 1991. *Computer Modelling in the Environmental Sciences.* Clarendon Press, Oxford.

Huggett, R., 1980. *Systems Analysis in Geography.* Clarendon Press, Oxford.

Lakhan, V.C. & Trenhaile, A.S. (eds), 1988. *Applications in Coastal Modelling*. Elsevier, Amsterdam.

Medley, D.G., 1982. *Mechanics and Modelling*. Heinemman, London.

Thornes, J.B. & Brunsden, D., 1977. *Geomorphology and Time*. Methuen, London.

Woldenberg, M.J. (ed.), 1985. *Models in Geomorphology*. Allen and Unwin, London.

Part III
EXAMPLES OF ENVIRONMENTAL MODELS

11
An Ocean Temperature Model

INTRODUCTION

The preceding chapters have provided an introduction to the philosphical background for computerised environmental modelling and to the construction, running and testing of a range of computer-based environmental models. These are the tools of our subject. Chapters 11–16 attempt to demonstrate what can be achieved with these tools and a little time. We describe some more complex examples of computer modelling together with suggested exercises and example outputs. We begin with a short chapter on a long subject: the modelling of fluid properties in the world ocean.

There is, within the environmental sciences, a whole group of models based upon full or partial differential equations because the environmental sciences deal, above all else, with either spatial or temporal gradients in system variables (see Chapters 1 and 2), and gradients or rates of change are best expressed within the language of the calculus. However, except for the most trivial of applications, environmental processes are non-linear (that is to say exponents other than unity are present in the governing equations; *cf*. Chapter 5) and linear differential equations are the only ones for which a complete analytical theory exists and for which general analytical solutions can be obtained (see Thornes & Brunsden, 1977,

p. 140 *et seq.* for a reasonable introduction to the use of differential equations in environmental modelling). There are, however, various techniques which can be applied to obtain satisfactory solutions to the equations, and the present and following chapters deal with one such technique: the use of finite element or finite difference analysis which, essentially assumes that, rather than a function being continuous, a particular solution exists over a region of space or time through grids of cells. In this way the equations may be solved and this approach is, in fact, the basis of much numerical modelling. In the present chapter, a model for temperature variation in the North Atlantic Ocean is described using a second-order diffusion equation and, in the following chapter, the same approach is applied to the dispersion of pollutants within a small river channel.

OCEAN MODELLING

The numerical modelling of water movements in the ocean's basins or in the world ocean as a whole or, indeed, of aspects of the atmosphere for meteorological prediction or the analysis of climatic change is a truly massive task, demanding the use of supercomputers and as yet unheard of processing requirements. Webb (1991), for example, notes that a reasonable grid resolution in a global ocean model requires computing power equivalent to many thousands of the most powerful of the current generation of desk-top machines. In particular, attempts to develop realistic models of the ocean circulation have been hampered for many years by the scale of important ocean features. In contrast to the atmosphere, where the major high and low pressure regions have scales of a thousand or more kilometres, the major ocean eddies have diameters of 100–200 km. Similarly, the major winds of the atmosphere, called the jet streams, may be hundreds of kilometres wide whilst in the ocean the major currents, like the Gulf Stream, are only 30–50 km wide. Nevertheless, reasonable models are now being constructed and are fulfilling a useful role within the ocean sciences. In the United States the US Community Model (Bryan & Holland, 1990) covers the North Atlantic between 15°S and 65°N, with a north–south resolution of 1/3°, and an east–west resolution of 2/5° and with 30 levels in the vertical. The global model of Semtner and Chervin (1988) uses a $1/2°$ grid in both the north–south and east–west directions with 20 levels in the vertical. In the United Kingdom, quite separate models have been developed for the prediction of tidal storm surges (Procter & Flather, 1983; Flather, 1984; Flather, Procter & Wolf, 1991) and for waves on the north-west European Continental Shelf and are described in Hardisty (1990). Additionally the Fine Resolution Antarctic Model (FRAM, Webb, 1991) covers the Southern Ocean and is being developed by collaboration between British institutes and university research groups.

In the present chapter, a much simplified North Atlantic model is described which, despite its rather limited scope, exemplifies the principles involved in ocean modelling and provides an interesting training platform.

THEORY

The code for the FRAM model (Webb, 1991), like those of Bryan and Holland (1990) and Semtner and Chervin (1988) is based on that developed by Bryan (1969), Semtner (1974) and Cox (1984). It splits the ocean into a grid along lines of latitude, longitude and depth and sets up arrays containing the temperature, salinity and horizontal velocity at each grid point. In essence, only three equations are necessary and these represent the continuity of thermal energy (i.e. the temperature of the water), the continuity of salt and the continuity of water. The equation for ocean temperature is:

$$\frac{dT}{dt} + u \cdot \nabla T + w \frac{dT}{dz} = A_h \nabla^2 T + K_h \frac{d^2 T}{dz^2} \tag{11.1}$$

where T is temperature, t is time, u and w are the horizontal and vertical velocities, A_h and K_h are the horizontal and vertical diffusion coefficients, and z is the vertical direction. The vector operator, ∇, is simply shorthand for the differential of a property and is called 'del'. Thus ∇T is the gradient or rate of change of temperature in the x, y and z directions (i.e. east–west, north–south and vertically) and the 'dot product' $u \cdot \nabla T$ is the rate of change of temperature along the direction of the horizontal current vector, u. A reasonable introduction to this type of vector analysis is given in the student edition of Spiegel (1959). The model developed here covers only temperatures and we shall, therefore, concentrate on this equation and simply note that the other two required for a full ocean model are the continuity of salt (S):

$$\frac{dS}{dt} + u \cdot \nabla S + w\frac{dS}{dz} = A_h\nabla^2 S + K_h\frac{d^2 S}{dz^2} \tag{11.2}$$

and the continuity of water (M):

$$\frac{dM}{dt} + u \cdot \nabla M + w\frac{dM}{dz} = A_h\nabla^2 M + K_h\frac{d^2 M}{dz^2} \tag{11.3}$$

and the problem is closed by assuming that the ocean is incompressible. It is instructive to note that all three equations have a similar form.

Concentrating effort on the ocean temperature problem, it is necessary to simplify the system somewhat in order to implement an operational model of the North Atlantic within Excel on a microcomputer. Practically we shall make the following assumptions.

- *Assumption 1.* The ocean is of uniform depth and homogenously mixed in the vertical sense. The vertical velocity and vertical diffusion terms can, therefore, be ignored in equation (11.1).
- *Assumption 2.* The heat input is only from the Equator, and heat is neither lost nor gained through either the ocean surface or the ocean floor. The model can then be cast as a single layer.
- *Assumption 3.* The ocean is stationary. The horizontal current terms can, therefore, be ignored in equation (11.1).

The results of these three assumptions are that equation (11.1) reduces to:

$$\frac{dT}{dt} = A_h\nabla^2 T \tag{11.4}$$

which, expanding the grad operator, becomes:

$$\frac{dT}{dt} = A_h \frac{\delta^2 T}{\delta x^2} + A_h \frac{\delta^2 T}{\delta y^2} \tag{11.5}$$

We note that this is the form of the diffusion equation used in the analysis of temperature waves passing through a glacier by Thornes and Brunsden (1977, p. 148). Physically it represents a change in temperature through time which is proportional to the rate of change of the temperature gradient (i.e. the second differential) in the east–west (x) and north–south (y) directions.

EXCEL IMPLEMENTATION OF THE CANAL MODEL

Two models are introduced in this section. The non-linear form of equation (11.5) is at once both interesting and complicated. As an example, consider that the North Atlantic from the Equator to 60°N (approximately the latitude of Newfoundland) may be represented by a canal of uniform cross section which is vertically and laterally homogeneous. We are reducing the problem to the type of one-dimensional models described in earlier chapters. The model is called simply 'Chapter Eleven' and is shown in Figure 11.1. The implementation consists of a control row (row A) within which the three input variables are set. Thus, an 'equatorial temperature', an 'initial northern temperature' and a 'diffusion coefficient' (i.e. A_h in equation (11.5)) are set. Rows 2–4 then contain the spatial variables, that is the latitudinal ranges of the 12 'cells' taken to define the canal, each 5° from 0° to 60°. Each cell is approximately 300 km long and thus row 3 contains the distance from the Equator given by the Excel formula:

$$= (300 * (\text{COLUMN}() - 2)) + 150$$

Row 4 contains an initial temperature distribution set by a function to decrease from the equatorial temperature in cell \$D\$1 to the northern temperature in cell \$I\$1. This function is of course arbitrary and could be changed. The resulting temperature variation at the

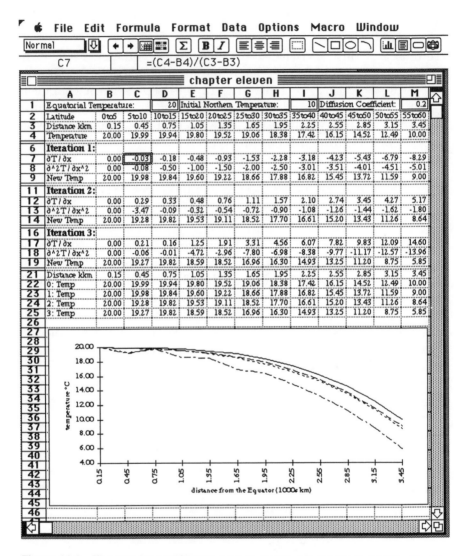

Figure 11.1 *The canal model for temperature changes in the North Atlantic showing embedded graphic plots of the original and subsequent temperature distributions*

beginning of the iterations is shown by the upper, solid line in the chart. It is essentially a cubic decrease with distance.

The model then performs three iterations using the diffusion equation. In each iteration the first differential of the temperature function is determined (i.e. row 7 and the formula in the formula bar in Figure 11.1), the second differential is determined (i.e. row 8) and then the temperature change is determined by multiplying the second differential by the diffusion coefficient contained in cell M1. The change is added to or subtracted from the previous temperature for each cell. The procedure is repeated three times and then the distances and four sets of temperatures are collected together in rows 21–25 and plotted in the chart.

It is clearly the initial temperature distribution and hence the local values of $\partial T/\partial x$ and $\partial^2 T/\partial x^2$ which control the model. These are plotted in Figure 11.2 for the conditions shown in Figure 11.1. The first differential (the broken line) is negative because the temperature is decreasing with distance from the equator. The second differential (the solid line) is also negative, meaning that the temperature gradient is becoming steeper with distance, but is a straight line,

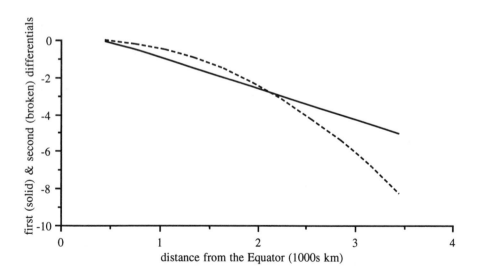

Figure 11.2   *Differentials plotted with distance from the Equator*

meaning that the gradient of the first differential is changing at a fixed rate. The result of equation (11.5) is therefore to decrease the ocean temperature at a rate which increases with distance from the Equator as was shown in the embedded chart.

RUNNING THE CANAL MODEL

Interesting use may be made of Excel's split-screen facility by dragging down the bar on the right-hand size box to display both the control modes and the chart for the canal model. Various 'what if' scenarios can then be examined because embedded charts respond automatically to spreadsheet changes. Thus, for example, setting the equatorial temperature to a value less than the northern temperature whilst keeping the diffusion coefficient constant produces the results shown in Figure 11.3. Here we see that the temperatures increase

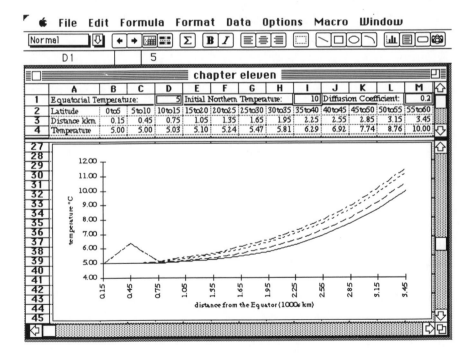

Figure 11.3 *Temperature plotted with distance from the Equator*

with time but at a decreasing rate. There is clearly a contrast here, since the first results showed a positive feedback, whilst the present results show a negative feedback in ocean temperature evolution.

EXCEL IMPLEMENTATION OF A LINEARISED SOLUTION

The non-linear form of equation (11.5) which was used to produce the canal model in the preceding section can be applied in both the north–south and east–west directions to generate a North Atlantic Ocean model, but these are beyond the scope of the present book. An alternative, and relatively interesting solution is to linearise the equations so that temperature changes depend simply on the first-order temperature gradient. The result is shown in Figure 11.4 and requires some explanation.

The model is based upon a grid of the North Atlantic Ocean with 15 columns each representing five degrees of longitude from 0° at Greenwich to 75° west, and 20 rows each representing three degrees of latitude from the Equator to 60° north. Column and row headings and numbers have been removed using the display sub-menu in the options menu. The top row is again the control and allows equatorial and mean temperatures and the diffusivity coefficient to be altered and set. The rest of the model consists of two grids; the lower contains the initial conditions and the upper the results of a single iteration. Cells were classified as land in the lower grid and shaded black using the patterns sub-menu in the format menu in accordance with the US Hydrographic Office World Charts sheet 7 and in general a cell containing less than 50% of ocean was classified as land. The resulting map is shown in the figure and includes north-eastern North America and Canada, north-eastern South America, Europe and North Africa.

Temperature evolution was modelled as a wave propogating northwards from the Equator, so that the temperature in each cell was set equal to the initial temperature plus a quarter of the temperature difference from the cells to the south-west and south-east and a half the difference from the cell to the south as shown by the formula bar in the figure. In practice the formulae were set up from the

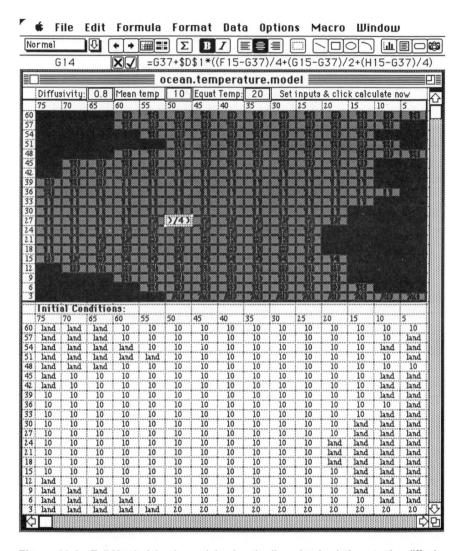

Figure 11.4 *Full North Atlantic model using the linearised solutions to the diffusion equation*

top left-hand corner and filled down and right, and then land cells and marginal cells were corrected as shown in the formula bar in Figure 11.5.

The model is run by setting the required inputs in the control panels and then choosing 'calculate now' in the options menu and a whole range of temperature evolutions can be investigated.

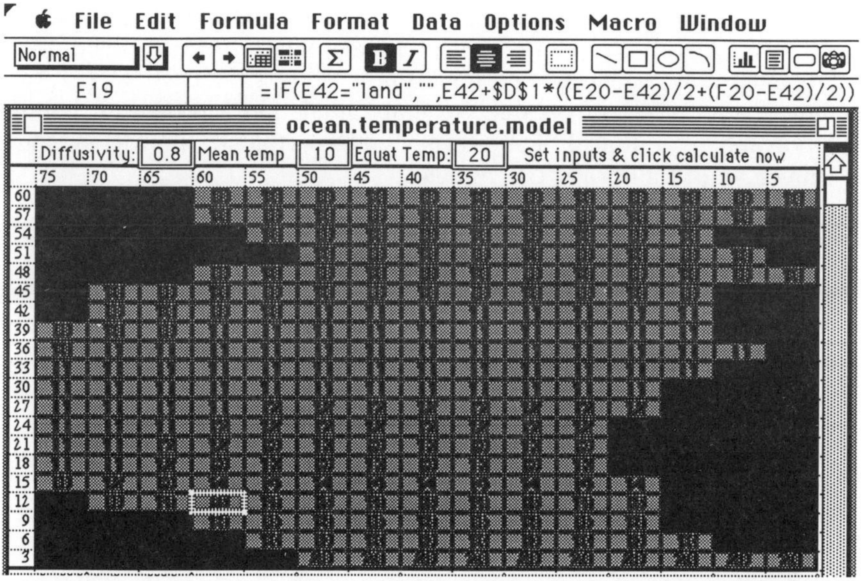

Figure 11.5 *Showing formulae correction for land in marginal cells*

REFERENCES

Bryan, F.O., 1969. A numerical method for studying the circulation of the world ocean. *Journal of Computational Physics*, **4**, 347–76.

Bryan, F.O. & Holland, W.R., 1990. A High Resolution Simulation of the Wind- and Thermohaline Driven Circulation in the North Atlantic Ocean. National Centre for Atmospheric Research, PO Box 3000, Boulder, Colorado 80307, USA (unpublished manuscript).

Cox, M.D., 1984. GFDL Ocean Group Technical Report No.1, Geophysical Fluid Dynamics Laboratory, Princeton University, Princeton, New Jersey 08542. USA.

Flather, R.A., 1984. A tidal model of the north-west European continental shelf. *Mémoires Société Royale des Sciences de Liege*, **10**, 141–164.

Flather, R.A., Procter, R. and Wolf, J., 1991. Oceanographic forecast models. In: D.G.Farmer & M.J.Rycroft (eds) *Computer Modelling in the Environmental Sciences*. Clarendon Press, Oxford, pp. 15–30.

Hardisty, J., 1990. *British Seas: An Introduction to the Oceanography and Resources of the North-west European Continental Shelf.* Routledge, London.

Procter, R. & Flather, R.A., 1983. Routine storm surge forcasting using numerical models: procedures and computer programs for use on the CDC Cyber 205 at the British Meteorological Office. Institute of Oceanographic Sciences Report, No.167.

Semtner, A.J., 1974. An oceanic General Circulation Model with bottom topography. Technical Report No.9, Department of Meteorology, University of California, Los Angeles.

Semtner, A.J. & Chervin, R.M., 1988. *J. Geophys. Res.*, **93**, 15502–15522.

Spiegel, M.R., 1959. *Vector Analysis*. McGraw Hill, New York.

Thornes, J.B. & Brunsden, D., 1977. *Geomorphology and Time*. Methuen, London.

Webb, D.J., 1991. FRAM—the Finite Resolution Antarctic Model. In: D.G.Farmer and M.J.Rycroft (eds) *Computer Modelling in the Environmental Sciences*. Clarendon Press, Oxford, pp. 1–14.

12
Chaos

INTRODUCTION

There is considerable interest at the present time in the emerging science of chaos whereby fully deterministic relationships appear to lead to apparently random results. Nowhere is this interest more intense than in the environmental sciences because it is possible that all of the great natural systems are chaotic. Thus, meteorological and climatological modellers are finding chaotic behaviours within the atmosphere, fluid dynamicists are invoking chaos to model turbulence and sedimentologists are finding chaotic orders in rock structures and stratigraphy. This book would not, therefore, be complete without some attempt to reproduce the beauty and fascination of chaos within the spreadsheet environment. What follows is a very simple model which consists essentially of one constant and one variable and yet we shall see that the result demonstrates a whole range of natural looking phenomena. Thus, the model introduces population dynamics which re-appears in Chapters 13 and 14. It introduces frequency doubling, a fascinating phenomena now known to be apparent in all manner of environmental systems. It also introduces the idea of self similarity, a concept central to the movement from the time domain to the frequency domain in many research projects and, when applied to landscapes, increasingly being held up as the prime candidate for a whole new generation of geomorphological theories. And all of this with one constant and one variable.

The chapter is based on the charming and beautifully written and illustrated book by Hans Lauwerier which has recently been translated into English and is recommended to the interested reader (Lauwerier, 1987).

CHAOS MODELLING

It is impossible to separate an analysis of chaos from the principle of self similarity. Laewerier (1987) populates his introduction of the two concepts with anecdotal descriptions of the individuals involved. Thus, it appears that the famous Swiss mathematician, Jacob Bernoulli (1654–1705, who was responsible for basic hydrodynamic theory as well as the Jacobian functions beloved of oceanographers) became so obsessed with the self similarity of spiral curves such that all were identical but operated at a different scale that he had the words *'eadem mutata resurgo'* ('though transformed, I will rise again unchanged') inscribed on his tomb in the cathedral of Basel.

Two centuries later the Scottish biologist, Robert Brown (1773–1858) discovered a peculiar phenomenon which exemplified self similarity. Looking through his microscope at small particles floating in a liquid, he was struck by the fact that the particles made tiny, erratic and unpredictable movements. It later transpired that this movement, now known as Brownian motion, was due to molecules in the liquid making irregular thermal motions that become more vigorous as the temperature rises. The molecules continually bump against larger particles that can be seen through the microscope. Brown could only see the larger scale motion but, with a microscope ten or a hundred times more powerful he would have seen virtually the same thing, and yet at all scales the motions appear random and chaotic. Following the work of fractal pioneers such as the Pole Benoit Mandelbrot and the Frenchmen Gaston Julia and Henri Poincaré we now know that the path of the Brownian particle is a fractal curve and there are analogous Brownian surfaces and Brownian landscapes. Many beautiful examples are illustrated in Mandelbrot (1982) in which Richard Voss used powerful computers and video techniques to generate quite artificial but very realistic images of a fractal landscape. These must rank as perhaps the most elegant computerised environmental models which have so far been constructed.

It was these general ideas that fully deterministic (or linear) processes can lead to apparently chaotic behaviour and that within these behaviours existed various sets of frequencies and attractors that led, in the 1970s and 1980s, to powerful insights into natural systems (see also Chapters 1 and 9).

THEORY

The model introduced here illustrates chaos and again concerns periodicities and self-similarities repeated to infinitely small scales. It was originally introduced as a model of unrestricted and restricted population growth which, for instance, describes the number of insects in successive generations (see also Chapters 13 and 14). We apply some scaling here so that the number always lies between 0 and 1. The model is:

$$x_t = ax_{t-1} \tag{12.1}$$

where x_t is the number in the current generation (or iteration) as a function of the number, x_{t-1}, in the last generation. There will thus be a times as many insects in any generation as in the generation before. The well-known economist, Malthus (1766–1834), who devoted much of his time to studying models of growth, is remembered in the name of the factor a which is nowadays called the Malthusian factor. In 1845 P. F. Verhulst extended the idea to consider examples of restricted growth by supposing that the Malthusian factor decreases as the number x increases. Supposing that the biggest (scaled) population that the environment will support is $x=1$ then if there are x insects, $1-x$ is a measure of the space permitted for population growth. Consequently we replace a by $a(1-x)$ in equation (12.1) and the restricted growth model becomes:

$$x_t = ax_{t-1}(1 - x_{t-1}) \tag{12.2}$$

We proceed by noting that this is a fully deterministic relationship, i.e. it is perfectly possible to predict x_t providing that a and x_{t-1} are specified and we should always get the same output for the same inputs. However, implementing the model in Excel vividly demonstates that we have defined a very chaotic system.

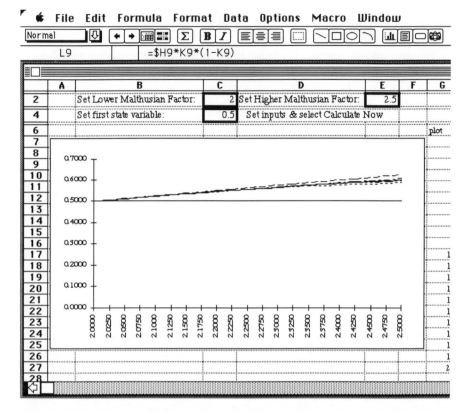

Figure 12.1 *Mode map of the Chaos model consisting of an embedded chart plotting the Malthusian factor on the x-axis against the value of the result from equation (12.2) for the first 20 iterations at each value of the factor. The array to the right (i.e. columns H to X) represents the iterations and rows 8 to 27 the factor values scaled to lie between the limits defined in the command boxes at top left. The first state variable (i.e. x at n=0, the beginning of the iterations) appear in column I and are also defined in the command boxes*

EXCEL IMPLEMENTATION

The mode map of the model (Figure 12.1) shows that the main display screen is headed with two control rows in which the initial state variable (i.e. $x_{t=0}$) and the required range of Malthusian factor values (i.e. a) are set. Columns H to X then represent 20 iterations (or generations) of the model in rows 8 to 27 for a values in the defined range. The Excel implementation of equation (12.2) is shown in the formula bar in Figure 12.1 and is, for cell L9:

G	H	I	J	K	L	M	N	O	P	Q	R	S	T
	n:	1	2	3	4	5	6	7	8	9	10	11	12
plot	a												
0	2.0000	0.5000	0.5000	0.5000	0.5000	0.5000	0.5000	0.5000	0.5000	0.5000	0.5000	0.5000	0.5000
1	2.0250	0.5000	0.5063	0.5062	0.5062	0.5062	0.5062	0.5062	0.5062	0.5062	0.5062	0.5062	0.5062
2	2.0500	0.5000	0.5125	0.5122	0.5122	0.5122	0.5122	0.5122	0.5122	0.5122	0.5122	0.5122	0.5122
3	2.0750	0.5000	0.5188	0.5180	0.5181	0.5181	0.5181	0.5181	0.5181	0.5181	0.5181	0.5181	0.5181
4	2.1000	0.5000	0.5250	0.5237	0.5238	0.5238	0.5238	0.5238	0.5238	0.5238	0.5238	0.5238	0.5238
5	2.1250	0.5000	0.5313	0.5292	0.5294	0.5294	0.5294	0.5294	0.5294	0.5294	0.5294	0.5294	0.5294
6	2.1500	0.5000	0.5375	0.5345	0.5349	0.5349	0.5349	0.5349	0.5349	0.5349	0.5349	0.5349	0.5349
7	2.1750	0.5000	0.5438	0.5396	0.5403	0.5402	0.5402	0.5402	0.5402	0.5402	0.5402	0.5402	0.5402
8	2.2000	0.5000	0.5500	0.5445	0.5456	0.5454	0.5455	0.5455	0.5455	0.5455	0.5455	0.5455	0.5455
9	2.2250	0.5000	0.5563	0.5492	0.5509	0.5505	0.5506	0.5506	0.5506	0.5506	0.5506	0.5506	0.5506
10	2.2500	0.5000	0.5625	0.5537	0.5560	0.5554	0.5556	0.5555	0.5556	0.5556	0.5556	0.5556	0.5556
11	2.2750	0.5000	0.5688	0.5580	0.5611	0.5603	0.5605	0.5604	0.5604	0.5604	0.5604	0.5604	0.5604
12	2.3000	0.5000	0.5750	0.5621	0.5661	0.5649	0.5653	0.5652	0.5652	0.5652	0.5652	0.5652	0.5652
13	2.3250	0.5000	0.5813	0.5659	0.5712	0.5695	0.5700	0.5698	0.5699	0.5699	0.5699	0.5699	0.5699
14	2.3500	0.5000	0.5875	0.5695	0.5761	0.5739	0.5747	0.5744	0.5745	0.5745	0.5745	0.5745	0.5745
15	2.3750	0.5000	0.5938	0.5729	0.5811	0.5781	0.5793	0.5788	0.5790	0.5789	0.5790	0.5789	0.5789
16	2.4000	0.5000	0.6000	0.5760	0.5861	0.5822	0.5838	0.5832	0.5834	0.5833	0.5833	0.5833	0.5833
17	2.4250	0.5000	0.6063	0.5789	0.5912	0.5861	0.5883	0.5874	0.5877	0.5876	0.5876	0.5876	0.5876
18	2.4500	0.5000	0.6125	0.5815	0.5962	0.5898	0.5927	0.5914	0.5920	0.5918	0.5919	0.5918	0.5918
19	2.4750	0.5000	0.6188	0.5838	0.6013	0.5933	0.5972	0.5954	0.5962	0.5958	0.5960	0.5959	0.5960
20	2.5000	0.5000	0.6250	0.5859	0.6065	0.5966	0.6017	0.5992	0.6004	0.5998	0.6001	0.5999	0.6000

Figure 12.1 (continued)

$$= \$H9 * K9 * (1 - K9)$$

where, for this cell, $\$H9$ is the value of the Malthusian Factor and K9 is the preceding value of the number (i.e. x_{t-1}). The results are displayed on the embedded chart which updates every time the model is run. The model is set to 'Manual Calculate' on the option menu and the column and row headers have been removed from the display sub-menu under 'format'.

RUNNING THE MODEL

The model can now be used to investigate chaos. Set the lower Malthusian factor to 2 and the upper Malthusian factor to 3.8 and click the 'Calculate Now' command in the option menu (this simply

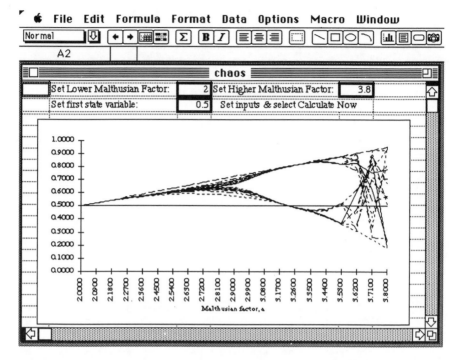

Figure 12.2 *Chaos model with upper and lower Malthusian factor values set to 2 and 3.8 respectively and an initial value of 0.5. Note the three regions which are identified in the text*

allows you to fiddle around with the inputs without the model constantly updating itself). The result is shown in Figure 12.2 and three distinct regions can be identified.

- *Single limit point region* Within the region $a = 2$ to $a = 2.5$ all iterations are returning the same value for the number and the system is in equilibrium. This region corresponds to convergence on a single attractor.
- *Period doubling region* Within the region $a = 3.1$ to $a = 3.3$ there are two equilibrium values returned by the iterations and the system is skipping between these two values. There is a transition between the first two regions and we shall examine this transition in more detail in a short while. This region corresponds to oscillations between two (and later four and eight and sixteen) stable states.

- *Chaotic region* Within the region $a > 3.65$ the system has become wholly unpredictable and each iteration is returning a different value. The system has become fully chaotic. There is again a transition region between the period doubling and chaotic behaviour.

Now investigate the self similarity of chaos. Change the lower and upper Malthusian limits to 3.35 and 3.6 respectively and double click on the embedded chart and then on the vertical axis and choose the scale option on the display box to limit the display as shown in Figure 12.3. Return to the model and run the new settings and the result is shown in Figure 12.4. The pattern appears to be disconcertingly similar to the one revealed earlier: there are single limit, period doubling and chaotic regions revealed at this new scale

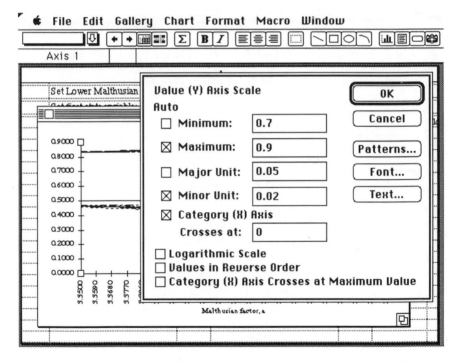

Figure 12.3 *Rescaling the vertical axis on the chart by double clicking on the chart and then clicking on the vertical axis and redefining the minimum and maximum values on the display*

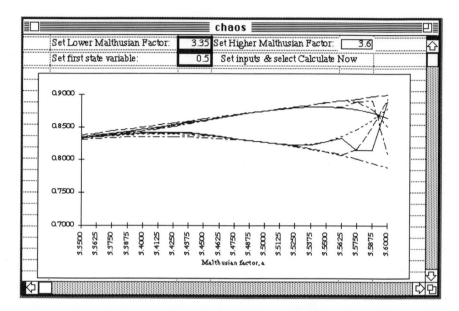

Figure 12.4   *Investigation of the smaller scale structure between Malthusian factor limits of 3.35 and 3.60 showing the same patterns of single point attractors, period doubling and chaos as in Figure 12.2*

within which the system shows remarkable self similarity. The model has been set up so that this can be examined in infinitely small detail, and order in chaos is revealed.

In an early paper Thornes (1983) begins to place geomorphology within this world of strange attractors and chaotic behaviour, and later (Thornes, 1987) he uses the same equation as described above. He notes that there are many examples in the natural environment of these quite sudden changes from stability to incoherence and cites the well-known transition from laminar to turbulent flow. The transition was well demonstrated in the now classical experiments of Osborne Reynolds who, in the late nineteenth century, set up the apparatus shown in Figure 12.5 with a valve controlling the discharge of water from a resevoir along a clear tube. Water movements within the tube were traced with a dye line. At low flow rates, the dye diffused slowly (and actually by Brownian motion (Figure 12.5(a)). At higher flow rates as the valve was opened the dye streak began

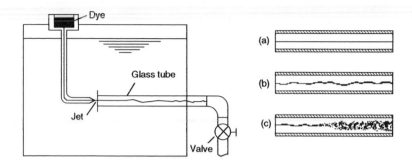

Figure 12.5 *The Reynolds experiments illustrating the sudden change from ordered, laminar flow (a) to chaotic turbulent flow (c) (from Hardisty, 1993)*

to oscillate (Figure 12.5(b)) and then, with a small further increase in the flow rate, the dye streak became immediately mixed across the whole tube as the flow became chaotic and turbulent (Figure 12.5(c)). In a third paper, Thornes (1990) remarks on the self-similarities of landforms and begins to perceive a Brownian Cascade from large mountains through hills and hillocks to small ephemeral features controlled by chaotic order within a random world.

REFERENCES

Hardisty, J. *Sediment Dynamics in Rivers, Deserts and the Marine Environment.* Blackwells, Oxford, 1993.

Lauwerier, H. 1987. *Fractals, Endlessly Repeated Geometrical Figures.* Penguin Books, London. (Translated by Sophia Gill-Hoffdstadt, 1991).

Mandelbrot, B.B., 1982. *The Fractal Geometry of Nature.* W.H.Freeman, London.

Thornes, J.B., 1983. Evolutionary geomorphology. *Geography,* **68**, 225–235.

Thornes, J.B., 1987. Environmental systems. In: M.J. Clark, K.J.Gregory & A.M.Gurnell (eds) *Horizons in Physical Geography.* Macmillan Education, London, pp. 27–46.

Thornes, J.B., 1990. Big rills have little rills.... *Nature,* **345**, 764–765.

13
Daisy World

INTRODUCTION

The modelling of environmental feedback mechanisms was introduced in Chapter 9 with the example of positive and negative feedbacks and hillslope evolution. In climatic systems, feedback occurs when a product of a climatic process (for example temperature at the Earth's surface) has some effect on the climatic process (for example atmospheric circulation): positive feedback causes an amplification of the perturbation, whilst negative feedback results in a dampening. Feedback mechanisms which compensate for disrupting change are also common in biology, where they are known as homeostatic processes. 'DAISY WORLD', the model developed in this chapter, is a very simple example of a biospheric feedback model and was originally constructed by supporters of the Gaian hypothesis to illustrate a way in which biotic activity could influence the abiotic (or non-biological) environment (in this case climate). In addition to basic climatic modelling, DAISY WORLD therefore also serves as an introduction to Gaian philosophy. As will become apparent later, it also provides an opportunity to test if there is a relationship between system stability and diversity.

GAIAN PHILOSOPHY

The biosphere is the 'envelope of life' on Earth. It can be divided into two components: the biotic (living organisms, or biota, and their remains) and the abiotic (the relevant inorganic parts of the atmosphere, geosphere and hydrosphere). There is some degree of interrelationship between the two. For example, few would argue that the state of the abiotic environment has no influence over the structure and composition of a biota. However, the extent to which the reverse is true (i.e. the degree to which the abiotic environment is dependent upon the biota) remains the subject of some debate.

Gaian philosophy (from Gaia, the Greek name for the Earth goddess) argues that the abiotic and biotic components of the environment are strongly intertwined, and that the composition and structure of the biota are not simply a reflection of physical conditions, but actually play an important role in determining and maintaining the abiotic environment. Rather than the development of one component following the other, the biosphere is therefore perceived as evolving as a single entity. In other words, the nature of the Earth's atmosphere and surface is actively regulated by, and regulates, the sum of life on the planet. As an example, Lovelock (1988, 1989), an early proponent of Gaian philosophy, cites the difference in atmospheric composition between a lifeless planet, such as Mars or Venus, and one supporting life, i.e. Earth. On the lifeless planets carbon dioxide forms a major part of the atmosphere whereas it is relatively rare in the Earth's atmosphere. Indeed the level of carbon dioxide in the Earth's atmosphere is found to be well below the expected chemical equilibrium (about 30 times less), a situation which Lovelock suggests could only have arisen as a consequence of biological activity. Carbon dioxide is not the only atmospheric gas with a level appearing to be at least partially biologically controlled; the levels of gases such as ammonia, nitrous oxide, sulphate and formaldehyde in the Earth's atmosphere are very different to the equilibrium or steady state and different to those observed or expected for planets without a biota. In other words the Earth's atmosphere is more than just part of the physical environment as its composition is at least partly biogenic. Lovelock goes on to suggest that the biotic influence is not incidental, nor is it merely restricted to the atmosphere. Instead he suggests that the biota maintain their environment in a form highly suitable (i.e. at the 'optimum') for life and that a biotic influence is equally strong in the hydrosphere (e.g. in controlling salinity of the oceans) and in the geosphere (e.g. determining the nature of many sedimentary rocks).

Fundamental to the Gaian view is that the biota actively engage in environmental regulation and control *for their own good*. Thus the biosphere can be regarded as a *cybernetic system*, in other words one which is steered towards a certain range of environmental conditions (or 'goal') that is the optimum for life. Once this has been achieved, Gaian philosophy holds, the conditions are maintained by the biota through homeostatic control. The suggestion that the biosphere is a directed system that, once at the optimum state for life, is maintained by negative feedback loops has been criticised by a number of scientists (e.g. Doolittle, 1981; Dawkins, 1982; Kirchner, 1989). Notwithstanding problems over the definition of 'optimum conditions for life' (there is no such thing—the species which go to make up the biota have different requirements and therefore different optimum conditions), there is evidence to indicate that the biosphere does not behave as a system dominated by negative feedbacks. For example followers of Gaia have proposed the biogenic production of carbon dioxide and cloud condensation nuclei (CCN) as mechanisms through which the biota regulate climate (Charlson *et al.*, 1987). However, this is not supported in palaeoatmospheric and palaeoclimatic records, such as the one provided by the Vostock ice core (Barnola *et al.*, 1987; Jouzel *et al.*, 1987): rather than playing a moderating homoeostatic role, biotic activity appears to accentuate climatic change, causing the climate to get colder when cold and warmer when warm (Legrand, Delmas & Charlson, 1988).

In an effort to illustrate a mechanism by which the biota might optimise their abiotic environment through negative feedback, Watson and Lovelock (1983) introduced DAISY WORLD. The model is *heuristic* in that it was constructed as a tool for instruction. In the case of DAISY WORLD the model was originally constructed to provide an example of how the biotic and abiotic components of the biosphere *could* be linked in a way that would facilitate some biotic control over the environment, and not to show how the two are *actually* linked.

DAISY WORLD

DAISY WORLD is an imaginary planet, with a transparent atmosphere, free from clouds and greenhouse gasses. The planet is flat, resulting in similar changes in temperature with changing solar luminosity (energy from the sun) and albedo being experienced simultaneously over its surface, and does not experience any seasonality in climate. The composition of the planet's biota is similarly lacking in complexity: two species of daisies occur as discrete populations; one dark (black), the other light (white) in colour. In addition a species of herbivore grazes the daisies in a non-selective manner (i.e. they show no preference for black or white daisies) and is responsible for the recycling of any organic material. The herbivores do not, however, exert any measurable effect on the system and are thus not further considered here.

Conditions on DAISY WORLD are suitable for the growth of daisies over the entire surface of the planet. Because of the difference in albedo, local temperatures are greater above the dark coloured daisies (which reflect less energy and are therefore warmer) than above the light coloured daisies. Apart form the difference in colour, both light and dark daisies are identical and have similar parabolic growth responses to temperature (with growth occurring between 5 °C and 40 °C and peaking at 22.5 °C). However, because of the difference in albedo and thus local temperature, the dark daisies have a faster growth rate at lower globally averaged temperatures than the lighter daisies. At higher globally averaged temperatures the reverse is true.

The rate of expansion of area covered by the two populations of daisies is, therefore, dependent upon the local temperature and amount of fertile land available, and upon the death rate. In the DAISY WORLD model the death rate is a constant, whereas local temperatures and the availability of fertile land are variables and are indirectly dependent upon the level of incoming solar radiation, which steadily increases with time. Construction and implementation of DAISY WORLD is described in the following sections. The objective is to illustrate the controlling influence a biota can have on their environment, so that optimum conditions for their growth are maintained over a longer time period than would be the case on a planet without life. Therefore two models need to be constructed and their output compared; one model with life and a second model without life.

THE DAISY WORLD MODEL

The model involves a number of formulae (the reader is advised to consult Henderson-Sellers & Robinson (1986) for a detailed explanation of the climatic terms referred to):

1. Firstly we need to determine the amount of fertile land available for growth:

$$x = [P - (a_b + a_w)]$$

   where

   x = amount of available land,
   P = proportion of land available for growth = 1.0,
   a_b = area of black daisies = 0.2 (initially) and
   a_w = area of white daisies = 0.2 (initially).

2. Next we need to determine the total (overall) albedo for the planet (albedo = the amount of radiation out divided by the amount of radiation in, i.e. it is usually less than 1.0):

$$A = x(Ag) + a_b(A_b) + a_w(A_w)$$

   where

   A = albedo of planet,
   Ag = albedo of bare ground = 0.5,
   A_b = albedo of black daisies = 0.25, and
   A_w = albedo of white daisies = 0.75.

3. We now need to determine the globally averaged temperature of the planet:

$$Te = \left(\frac{SL(1-A)}{s}\right)^{0.25} - 273$$

where

Te = globally average temperature,

S = a solar constant (energy from the sun) = 1000,

L = luminosity (proportion of present day value) = 0.7 initially but increasing in steps of 0.025, and

s = Stefan's constant = 5.67×10^{-8}.

4. The formula to work out the local temperatures for populations of black and white daisies (as these affect the growth rate) is:

$$T_{b,w} = (q(A - A_{b,w})) + Te)$$

where

T_b = local temperature of black daisies,

T_w = local temperature of white daisies, and

q = a constant used to calculate local temperature as a function of albedo = 20.

5. The growth rate of the populations of black and white daisies is determined from:

$$B_{b,w} = \{1 - [0.003265((22.5 - T_{b,w})^2)]\}$$

where

B_b = growth rate for black daisies,

B_w = growth rate for white daisies, and 1, 0.003265 and 22.5 are all constants so that growth occurs between 5 and 40 °C and peaks at 22.5 °C.

6. The change in area of black and white daisies is:

$$da_{b,w}/dt = \{a_{b,w}[(xB_{b,w}) - y)]\}$$

where

da_b = change in area of black daisies,

da_w = change in area of white daisies,

y = death rate = 0.2, and

t = time.

7. Hence the new area of black and white daisies is:

$$Na_{b,w} = ((da_{b,w}/dt) + a_{b,w})$$

where

Na_b = new area of black daisies, and

Na_w = new area of white daisies.

(*Note*: '*b,w*' means that the formula should be solved for both black and white daisies separately.)

EXCEL IMPLEMENTATION

The mode map of the model (Figure 13.1) shows the organisation of the parameters and their values. Solutions to the formulae to determine the changes in temperature and area occupied by the two populations of daisies are located in rows 14:23. The new areas occupied by white (B22) and black (B23) daisies after one 'growing season' should be carried forward into the next cycle of the model (i.e. taken up by cells C3 and C4 respectively).

Solar luminosity is the controlling variable in DAISY WORLD. The amount of solar luminosity received increases with time, as is the case with any ageing planet; thus with each cycle of the model the luminosity value (row 10) should be increased by 0.025 (i.e. the value of luminosity will be = 0.7 + 0.025 in the second cycle of the model,

$\bullet$ File Edit Formula Format Data Options Macro Win 1:50:47

| B15 | | =(B14*B7)+(B3*B5)+(B4*B6) | |

DAISYWORLD

	A	B	C
1	Model of Gaian control of climate		
2	Proportion of planet's area suitable for growth	1.00	1.00
3	Area covered by white daisies	0.20	0.12
4	Area covered by black daisies	0.20	0.24
5	Albedo of white daisies	0.75	0.75
6	Albedo of black daisies	0.25	0.25
7	Albedo of bare ground	0.50	0.50
8	Death rate	0.20	0.20
9	Solar constant	1000.00	1000.00
10	Luminosity	0.70	0.73
11	Stefan's constant	0.000000057	0.000000057
12	Constant used to calculate local temperature as a function of albedo	20.00	20.00
13			
14	Amount of fertile land available	0.60	0.64
15	Albedo of planet	0.50	0.47
16	Temperature of planet	7.30	13.89
17	Local temperature for white daisies	2.30	8.29
18	Local temperature for black daisies	12.30	18.29
19	Growth rate for white daisies	-0.33	0.34
20	Growth rate for black daisies	0.66	0.94
21			
22	New area white daisies (check >0)	0.12	0.12
23	New area black daisies (check >0)	0.24	0.34

Figure 13.1 *Worksheet for the DAISY WORLD model*

0.725 + 0.025 in the third cycle, etc). One point worth noting is that it is important to ensure that there are no negative areas of daisies in the model. An 'IF' statement should therefore be inserted in the program to check for negative areas so that, if any do occur, they are converted to a zero (0). The format and operation of IF statements are described in Chapter 8.

Using the information given, construct the model as in column B in Figure 13.1. To 'run' the model simply copy column B into columns to the right, making sure that luminosity is increased by 0.025 at each step and that the new values for the areas of white and black daisies are carried over into the next column, until the areas of both light and dark daisy populations have dropped to zero levels. The results of the model should be expressed in the form of two charts in which changes in luminosity are separately plotted against changes in globally averaged temperature and changes in the areal extent of light and dark daisies (Figures 13.2(a) and (b)). The first of these can then be compared with a similar plot for a planet without life (Figure 13.3), results for which can be obtained simply by changing the initial values of the a_b and a_w variables to zero (0) and re-running the model.

DAISY WORLD is obviously greatly simplified when compared to the real world. However, this can be addressed to some extent by the introduction of additional species of daisies, each with a different colour and hence albedo, as described in Lovelock (1988, 1989). Once you have familiarised yourself with the operation of the model as described here, construct your own, more complex, version in which, rather than just two species of daisies, DAISY WORLD is occupied by six species, each of which has a different albedo (but all of which are able to grow only between 5 and 40 °C and have peak growth rates at 22.5 °C). Run your model and chart increasing luminosity against temperature and against changes in the areas of the six species of daisy. This should allow you to comment on whether increased diversity (in this case an increased number of daisy species) leads to greater system stability (in this case the length of the period during which temperature is maintained at levels suitable for growth). Proving or refuting the theory of *increased diversity equals increased system stability* has taxed ecologists over many decades.

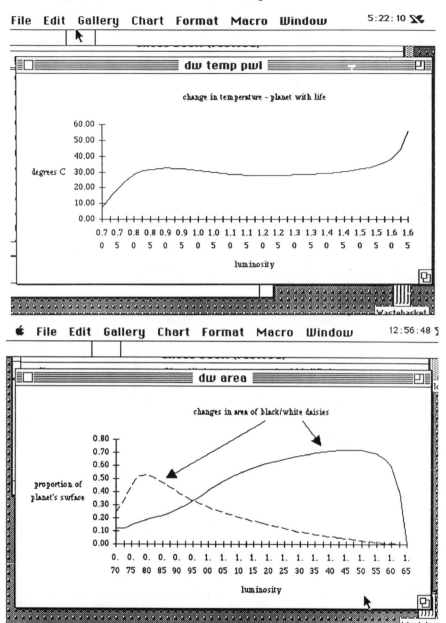

Figure 13.2   *(a) Change in temperature with time on a planet with life. (b) Changes in the proportion of fertile land occupied by separate populations of black and white daisies with increasing luminosity*

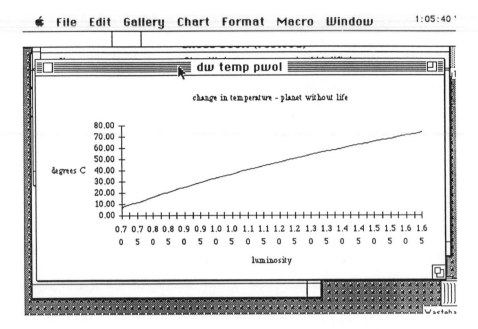

Figure 13.3 *Change in temperature with time on a planet without life*

REFERENCES

Barnola, J.M., Reynaud, D., Korotkevitch, Y.S. & Lorius, C., 1987. Vostock ice core provides 160,000-year record of atmospheric CO_2. *Nature*, **329**, 408–414.

Charlson, R.J., Lovelock, J.E., Andreae, M.O. & Warren, S.G., 1987. Oceanic phytoplankton, atmospheric sulfur, cloud albedo and climate. *Nature*, **326**, 655–661.

Dawkins, R., 1982. *The Extended Phenotype*. W.H. Freeman, Oxford.

Doolittle, W.F., 1981. Is nature really motherly? *Coevolution Quarterly*, **29**, 58–65.

Henderson-Sellers, A. & Robinson, P.J., 1986. *Contemporary Climatology*. Longman, Harlow.

Jouzel, J., Lorius, C, Petit, J., Genthon, C., Barkov, N., Kotlyakov, V. & Petrov, V., 1987. Vostock ice core: A continuous isotope temperature record over the last climatic cycle (160,000 years). *Nature*, **329**, 403–408.

Kirchner, J.W., 1989. The Gaia hypothesis: can it be tested? *Reviews of Geophysics*, **27 (2)**, 223–235.

Legrand, M.R., Delmas, R.J. & Charlson, R.J., 1988. Climatic forcing implications from Vostock ice core sulphate data. *Nature*, **334**, 418–420.

Lovelock, J.E., 1988. *The Ages of Gaia: A Biography of Our Living Earth*. Norton, New York.

Lovelock, J.E., 1989. Geophysiology, the science of Gaia. *Reviews of Geophysics*, **27 (2)**, 214–222.

Watson, A. & Lovelock, J.E., 1983. Biological homoeostasis of the global environment: the parable of the 'daisy world'. *Tellus*, **35B**, 284–289.

14

Faunal Extinction on an Isolated Island

INTRODUCTION

We are presently experiencing a 'diversity crisis' in which biological (i.e. genetic) diversity is experiencing a rapid decline as a result of high rates of species extinction. This is of obvious concern for conservationists and natural resource managers. It is also a source of concern amongst economists and politicians as species loss is likely to affect the future welfare of human populations as many of the species lost to date have, or might have had in the future if they had survived, an economic value. Furthermore habitat destruction, the main cause of extinctions at present, may have a wider impact on the environment such as through accentuating global climatic change.

The EXTINCTION model described here is based upon discrete time stages (i.e. the difference equations first mentioned in Chapter 1) and simulates changes in the population levels of an island's fauna over time, occurring as a result of an expansion (in both numbers and area occupied) of human predators. Depending on the initial values given to the various parameters used in the model, the fauna either happily coexist with their human predators or become extinct. It is assumed that if the latter takes place, in the absence of alternative sources of food, the predator population will itself face extinction if it is unable to evacuate the island. The latter may actually have taken place in the past on what Martin (1990) refers to as 'islands of doom'. These are islands, such as Henderson Island in the Pitcairn Group, which were uninhabited during the historical period but today show evidence of abandoned prehistoric human settlements associated with the remains of a now extinct fauna (Steadman & Olson, 1985).

PAST AND PRESENT DIVERSITY CRISES

The present diversity crisis is undoubtedly human-induced, both through the direct (e.g. over-hunting) and indirect (e.g. habitat destruction) actions of human populations. Less is known about the cause or causes of earlier periods of biotic impoverishment, which are apparent in the geological record as phases of 'mass extinction'. Unlike the present crisis, however, loss of species during these earlier extinction phases appears to have been compensated by the products of speciation (i.e. the evolution of new species).

Perhaps the best known examples of mass extinction in the geological past took place in the later stages of the Permian and Cretaceous periods approximately 220 and 70 million years ago, respectively. The Permian diversity crisis was possibly the most dramatic, in terms of the sheer numbers of species involved (an estimated 85–95 % of shallow marine species became extinct). However, palaeontologists now believe that the 'suddenness' of the event was more apparent than real, and that it occurred over a period of approximately 20 million years (van Andel, 1985). A second mass extinction event marks the Cretaceous–Tertiary boundary in geological strata, and may have taken place over a much shorter time period. It involved the extinction of both marine (e.g. ammonites) and terrestrial (e.g. dinosaurs) species, although not necessarily simultaneously, and its cause continues to be the focus of much debate (e.g. Alvarez *et al.*, 1980).

A more recent mass extinction phase occurred during the later stages of the last ice age (the Late Pleistocene), during which many species of large mammals (the so-called 'Pleistocene megafauna', e.g. the giant Irish elk and the mammoth) became extinct. Although other smaller animals, e.g. birds (Grayson, 1977), also disappeared, the Late Pleistocene extinctions are unique in that they mainly involved large, terrestrial mammal species. The cause or causes of these extinctions have been the subject of much discussion (e.g. Meltzer & Mead, 1983; Barnosky, 1986; Grayson, 1987; Martin, 1990). However, unlike earlier phases, we have a clearer idea of the overall environment in which the extinctions took place. This is largely because radiocarbon dating has provided an accurate chronology for the event and has thus allowed use to be made of relevant palaeoecological and palaeoclimatic data sets.

According to the radiocarbon dates so far obtained, the majority of Pleistocene megafaunal extinctions appear to have occurred roughly between 15 000 and 10 000 years ago. This is known to have been a period of relatively rapid environmental change at the end of which grasslands, the preferred habitat of many of the species involved, were much less extensive than before. Consequences of a reduced extent of suitable habitat would have been decreased population levels and increased isolation and inbreeding of surviving populations. Together these would have reduced the fitness of any surviving animals, thus rendering them more susceptible to extinction through catastrophes, such as disease. The period was also one of increased human activity and technological advancement, and the megafauna would have made an attractive target to groups of palaeolithic hunters (many megafauna fossils are associated with prehistoric butchering sites), whilst their inherently low rates of reproduction would have made recovery from over-hunting difficult. Furthermore, by concentrating populations into relatively small areas, climatic change may have indirectly favoured palaeolithic hunters. Hence the current consensus of opinion is that a combination of climatically-induced habitat destruction and palaeolithic overkill brought about the Late Pleistocene megafaunal extinctions (Goudie, 1981), with the relative importance of the two factors possibly exhibiting regional and taxonomic variations. The decline of those species which were not themselves the targets of hunters may have been a direct result of environmental change. The extinction of the giant Irish elk could be an example of this, as it appears to have disappeared from Ireland long before the development of a significant human population (Barnosky, 1986). In other regions, for example North America, where there is abundant evidence for the hunting of megafauna, human overkill is likely to have been a much more important phenomenon. In both cases species extinction may, through trophic cascade, have led to the extinction of other dependent species.

The Late Pleistocene phase of extinctions is, to a certain degree, analogous to the present diversity crisis. We are in, or about to enter, a period of rapid climatic change; with continued technological advance and projected increases in human population levels and requirements expected over the next 100 years or so, we are certainly in a period of rapidly expanding human influence. As was the case at the end of the last ice age, those species with a low fecundity and close dependence upon other species or upon a particular habitat will be the most vulnerable to extinction over the coming decades (Myers, 1990).

THE EXTINCTION MODEL

EXTINCTION is based upon the simulation of overkill by palaeoindian hunters described by Mosimann and Martin (1975). It was originally constructed to illustrate the potential rate at which a fauna could be destroyed by a rapidly expanding and unfamiliar predator. For convenience in modelling, EXTINCTION is based upon a geographically homogeneous island. The island is sufficiently isolated to prevent prey population levels from being supplemented by immigration and has the shape and area of a quarter circle of radius 400 km. Initially the human predators are established at the apex of the island, at which point the island's fauna is evenly distribution throughout the whole area (Figure 14.1(a)). Initial levels of prey and human population are respectively set equal to and equal to or below the environmental 'carrying capacity' (i.e. the maximum density the environment can support). Provided that a surplus of prey remains after predation, the level of human population increases geometrically and the occupied area is increased by a 'smooth' advance so that the population density at time $t+1$ is equal

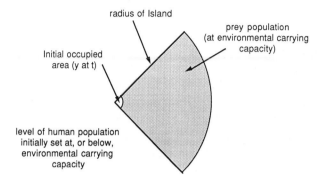

A. Island at time t

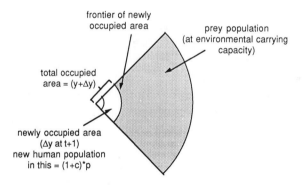

B. Island at time t+1
(after smooth or jump advance)

Figure 14.1 *Isolated and geographically homogenous island used in the EXTINCTION model (for explanations of symbols and formulae, see text)*

to that at t. In years of prey deficit (i.e. during years when predation exceeds the rate of prey replacement) no human population increase takes place and a 'jump' advance occurs. In this, the 'front' of the area occupied by humans is moved a fixed distance from the island's apex (Figure 14.1(b)). If insufficient prey is made available by one jump advance then further jump advances are made.

Once sufficient levels of prey are found, human population levels increase geometrically and any further advances will be smooth until the next year of prey deficit, or until the island is completely occupied. Once completely occupied, whether extinction takes place or not will depend upon the relative values for rates of prey destruction, prey replacement and human population increase, and upon the respective levels of environmental carrying capacity for predator and prey.

EXCEL IMPLEMENTATION

In addition to the normal Excel worksheet (Figure 14.2), EXTINCTION also includes three function macros (Figure 14.3). These are series of stored instructions which, once activated, use selected data to calculate a value. This value is then returned to a designated cell on the worksheet. The advantage of function macros is that they increase considerably the modelling power and speed of Excel. In order to create a function macro, select the new option from the file menu and then create a new Macro Sheet. To name the function macro, type an appropriate name in cell A1 of the Macro Sheet, select cell A1 and the define name option on the formula menu, select the function macro option from the dialogue box and choose 'OK'. The value calculated by the the function macro is returned to the worksheet using the command '=RETURN()'.

The function macros incorporated in the extinction model are Macro1!-Reproduction (called in row 20 of the Excel worksheet), Macro2!Smooth and Macro3!Jump (called in row 24). These respectively determine the prey biomass after reproduction at $t+1$; the extent of human occupation after a smooth advance as a result of a prey surplus; and the extent of human occupation after a jump advance, or series of jump advances, as a result of a prey deficit.

Normal

B25   =IF(B24="",Macro2!SMOOTH(B21,$B9,B11,$B13,$B15,
$B12,B17),Macro3!JUMP($B13,$B5,$B14,B17,B11,B20
$B12,B21,$B9))

1	modelling the decline of fauna, since			
2	appearance of human predators on an isolated island.			
3				
4	actual prey biomass (number of animals per km)	20.00	16.00	12.00
5	prey carrying capacity	20.00	20.00	20.00
6	prey replacement rate	0.25	0.25	0.25
7	number of human predators	20.00	25.00	31.25
8	maximum human growth rate	0.25	0.25	0.25
9	ceiling human density	2.00	2.00	2.00
10	prey destruction rate	4.00	4.00	4.00
11	distance of front from origin	3.57	3.99	4.46
12	Maximum distance front can move from origin	400.00	400.00	400.00
13	pie	3.14	3.14	3.14
14	increase in front from origin (=minimum jump)	25.00	25.00	25.00
15	maximum smooth advance	50.00	50.00	50.00
16				
17	original area occupied by humans	10.00	12.50	15.63
18	human density	2.00	2.00	2.00
19	prey biomass remaining after predation	12.00	8.00	4.00
20	level of prey after reproduction at end of year	15.00	10.00	5.00
21	new human population	25.00	31.25	39.06
22				

Figure 14.2   Worksheet for the EXTINCTION model

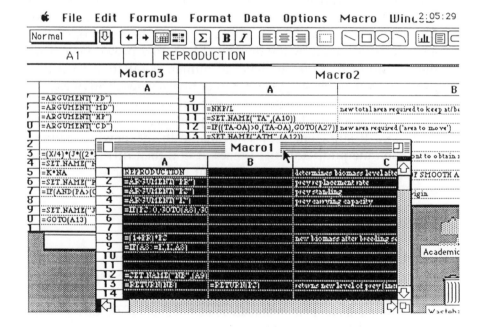

Figure 14.3   Macros for the EXTINCTION model

All three macros must be opened before any attempt is made to run EXTINCTION.

The EXTINCTION model proceeds as follows:

1 At time t, the density of human predators is equal to M/Y, where M is the number of people and Y is the area occupied.
2 The prey biomass remaining after predation thus equals $N - (G * M)$, where N is the prey biomass (originally the same as the environmental carrying capacity) and G is the destruction rate (number of animals killed per head of human population).
3 Provided that some prey remains, prey levels at $t+1$ will be equal to $(1 + A) * N$, where A is the rate of prey replacement.
4 With sufficient supplies of food (i.e. a prey surplus) the human population level will increase geometrically to equal $(1 + C) * P$ at $t + 1$, where C is equal to the rate of reproduction, and P is the population level at t (originally set at or below the carrying capacity).
5 In order to keep the density of human population below the carrying capacity, a smooth advance occurs (up to a maximum of 50 km) during which the area occupied by humans is increased. The new area of island occupied (ΔY) is thus the new population level at $t+1$ divided by the density at t. During prey deficit years there is no increase in the level of human population level (the population is under stress) and, rather than a smooth advance, the front of the occupied region is advanced in steps until sufficient prey is made available. The first jump advance covers a distance of 25 km. If sufficient prey is available in the newly occupied area, then the jump advance ceases. If not, then further jumps of 2 km are made until there is sufficient prey available.
6 In both smooth and jump advances,

$$R = (D^2 + (4 * \Delta Y/\pi))^{0.5} - D$$

where R is the distance of advance, D is the distance the radius of the occupied area at t. $D + R$ thus equals the radius of the occupied area at $t + 1$.

The colonisation process continues until there is no further new area on the island to occupy.

Model parameters are located in rows 4 to 15 and formulae (including references to function macros) in rows 17 to 31 on the Excel worksheet (Figure 14.2). Values in rows 5, 6, 8, 9, 11, 12, 13, 14 and 15 are constants. However, the values in rows 4 (actual prey biomass in occupied area), 7 (number of human predators in the occupied area) and 11 (distance of the front of the occupied area from the origin, or island's apex) are variables and are calculated by the formulae in rows 30, 21 and 27 respectively.

In order to run the model, open up the EXTINCTION worksheet and 'fill right' from column C (you will first need to ensure that the three function macros are open), until complete extinction of the island's fauna has taken place. Produce separate charts illustrating the increase in area occupied by human predators and the decline in total levels of fauna population (respectively Figures 14.4(a) and (b)).

According to the parameter values given in the version of

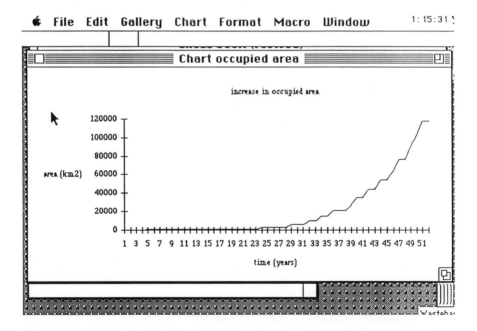

Figure 14.4 *(a) Chart to show increase through time in area occupied by humans. (b) (Overleaf) Chart to show decline in prey population through time*

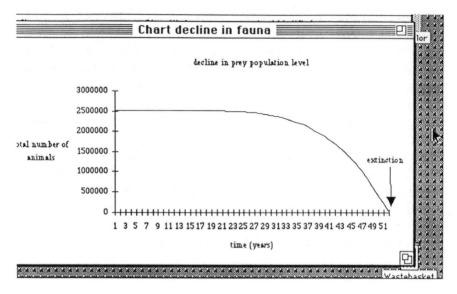

Figure 14.4   *(continued)*

EXTINCTION described here, the prey population becomes extinct on the island after 52 years. Having familiarised yourself with how the model operates, you might like to carry out some 'what if' tests of your own. For example, observe the effect of altering the parameters set in rows B4 to B15. Alternatively, you might like to have a go at solving a problem of *renewable resource management,* by determining the maximum yield of prey that is sustainable indefinitely (i.e. the highest rate of prey destruction that will not lead to the eventual extinction of the prey).

REFERENCES

Alvarez, L.W., Alvarez, W., Asaro, F. & Michel, H.V., 1980. Extraterrestrial cause for C - T extinction. *Science,* **208**, 1095–1108.

Barnosky, A.D., 1986. 'Big game' extinction caused by late Pleistocene climatic change: Irish Elk (*Megaloceros giganteus*) in Ireland. *Quaternary Research,* **25**, 128–135.

Goudie, A.S., 1981. *The Human Impact*. Blackwells, Oxford.

Grayson, D.K., 1977. Pleistocene avifaunas and the overkill hypothesis. *Science*, **195**, 691–697.

Grayson, D.K., 1987. An analysis of the chronology of late Pleistocene mammalian extinctions in North America. *Quaternary Research*, **28**, 281–289.

Martin, P.S., 1990. 40,000 years of extinction on the 'planet of doom'. *Palaeogeography, Palaeoclimatology, Palaeoecology*, **82**, 187–201.

Meltzer, D.J. & Mead, J.I., 1983. The timing of late Pleistocene mammalian extinctions in North America. *Quaternary Research*, **19**, 130–135.

Mosimann, J.E. & Martin, P.S., 1975. Simulating overkill by paleoindians. *American Scientist*, **63**, 304–313.

Myers, N., 1990. Mass extinctions: What can the past tell us about the present and the future? *Palaeogeography, Palaeoclimatology, Palaeoecology*, **82**, 175–185.

Steadman, D.W. & Olson, S.L., 1985. Bird remains from an archaeological site on Henderson Island, South Pacific: Man caused extinction on an 'uninhabited' island. *Proceedings of the National Academy USA*, **82**, 6191–6195.

van Andel, T.H., 1985. *New Views on an Old Planet*. Cambridge University Press, Cambridge.

15
Acid Deposition

INTRODUCTION

The acidification of surface waters and damage to buildings and vegetation have all been attributed to air pollution resulting largely from the combustion of fossil fuels, mainly the burning of coal in power stations (Nordic Council of Ministers, 1986). The full range of pollutants involved are sulphur species (SO_x), oxides of nitrogen (NO_x), oxidised nitrogen species (NO_y), ammonia, hydrocarbons and ozone. 'Acid deposition' is the term used for the removal of these pollutants by the processes of either wet or dry deposition. More commonly, this is called 'Acid Rain'.

Studies of long-term acidification of freshwaters in the UK have shown that the onset of acidification corresponded to the industrial revolution (Battarbee *et al.*, 1985). Increased levels of coal burning have been implicated. More recent sediments (from the 1940s on) contain carbonaceous particles produced by oil combustion. Power stations account for about 70% of UK SO_2 emissions, while NO_x emissions are derived from a wider range of sources including power stations, vehicle exhausts and agriculture (RGAR, 1990). Sulphur dioxide and nitrogen oxides are the main precursors of acid deposition.

Now that the link between fossil fuel combustion and

environmental damage (especially acidification of surface waters) has been clearly established, interest has become focused on strategies of cutting emissions to reduce this damage. Computer models have played a major role in the study of acid deposition and in assessing the effects of a range of future emissions scenarios.

MODELLING ACID DEPOSITION

A complete model would cover all processes from emission, through transport and chemical transformation, to deposition, catchment processes and the effects on ecosystems. Most models, however, focus on the stages from emission to deposition. Although models operate on a range of scales (spatial and temporal), long-range transport models have had the widest application in assessing relationships between source and receptor regions which may be thousands of kilometres apart. The range of modelling approaches available has been described by Hough and Eggleton (1986). Three main types of models are used: Lagrangian trajectory models (tracking the chemistry of a moving air parcel); Eulerian models (using a fixed grid co-ordinate system); and statistical models. Examples of these different types are respectively the models of: Eliassen and Saltbones (1983), a model for Europe; Chang *et al.* (1987), a model for the north-east USA and that of Fisher (1978), for UK/Europe. Most models deal only with sulphur because sulphur chemistry, transport and removal processes are quite well understood and emissions inventories are relatively easy to establish from large point sources.

The application of modelling studies in the UK has been reviewed in the Third Report of the Review Group on Acid Rain (RGAR, 1990). The three main models used are those developed at Harwell Laboratory (Lagrangian), by Warren Spring Laboratory (statistical) and by National Power (statistical). Although differing in detail, the models all reproduce the broad pattern of SO_2 concentration and of S deposition across the country. All the models tend to overestimate dry deposition and underestimate wet deposition in comparison with observations. A detailed comparison of output from the Harwell model with data from the UK Secondary Precipitation network, has been made by Metcalfe, Atkins and Derwent, (1989).

Using models, the processes of transport, chemical transformation and deposition can be explored and efforts made to attribute pollutants to different sources. In the UK for example, about 80% of total S deposition comes from indigenous sources, whereas in Switzerland >70% comes from foreign sources (Eliassen & Saltbones, 1983). The identification of source regions can assist in pollution control. The effects of emissions reductions, either already agreed (such as the Large Combustion Plant Directive (CEC, 1988)) or proposed, on pollutant concentrations or deposition, can also be explored using models. Using a Lagrangian model, a receptor point can be chosen and the changing contributions from different sources assessed as emissions are varied (Metcalfe & Derwent, 1990).

Recently, models have begun to include elements of more direct interest and concern to policy advisors, such as the costs of different abatement strategies (e.g. the ASAM model (Abatement Strategies Assessment Model) developed at Imperial College). One of the most widely used models in applied studies is that developed by IIASA (International Institute for Applied Systems Analysis). The RAINS model (Regional Acidification INformation and Simulation) (Alcamo *et al.*, 1987) deals with all aspects from pollutant generation to environmental impacts through a series of submodels, e.g. SO_2 emissions, cost analysis, lake acidification. Cost benefit analysis and models exploring the decision making process are also being developed.

THE MODEL

A very simple model of sulphur deposition is presented here using parameters derived from the Harwell model. The processes included are:

$$SO_2 \longrightarrow H_2SO_4$$

$$\uparrow \qquad\qquad \downarrow$$

$$\text{emission} \qquad \text{wet deposition}$$

The model follows the changing composition of a parcel of air along a straight line trajectory over a series of emission sources of differing magnitudes. Precipitation is assumed to be in the form of constant drizzle and the wind speed is set at 7.5 m s⁻¹. The screen display is shown in Figure 15.1.

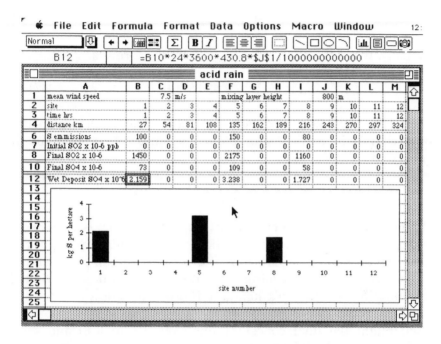

Figure 15.1 *Opening screen with initial SO$_2$ values set to zero*

The model consists of 12 sites which are located on a time scale at hourly intervals as shown in rows 2 and 3. The mean wind speed and mixing layer height are input as constants in cells C1 and J1 respectively. The mean wind speed is used to convert the time scale to a distance scale as shown in row 4. The sulphur emissions (in hundreds of metric tonnes per second) are entered into row 6. The initial values of the SO$_2$ concentration in each spatial cell is entered in row 7.

The model converts the S emissions into SO$_2$ in ppb (billion = 10^9) using the formula:

$$SO_2 = 1.45 \times 10^{-7}S \qquad (15.1)$$

and adds this amount to the 'initial' value in the preceding cell in row 8 which, therefore, represents a final SO_2 concentration. This concentration is converted into SO_4 in row 10 using:

$$SO_4 = 0.05SO_2 \qquad (15.2)$$

Finally, wet deposition, W_D, is given by the formula:

$$W_D = SO_4 \times 430.8 \times H_{mix} \qquad (15.3)$$

where H_{mix} is the height of the mixing layer. The result is displayed for each cell in row 12 and plotted against cell number in the embedded chart.

RUNNING THE MODEL

Although operating properly at this stage, this model is designed to be developed to permit the evolution of an equilibrium deposition rate to be observed. In particular, it is a relatively simple operation to use the 'final' concentrations after any iteration as 'initial' concentrations for the next iteration. Proceed as follows:

1 Highlight the final concentrations (cells B8 to M8) and select 'Copy' from the Edit menu.
2 Select the initial cells (B7 to M7) as shown in Figure 15.2.
3 Choose the 'paste special' option from the edit menu and then select 'values' in the dialogue box as shown in Figure 15.3. This means that only the values of the final concentrations will be pasted into the selected cells and not the formulae. If this is not done, then some erroneous values arise.
4 Click 'OK' and the new initial values form the start point for the next iteration. Repeat for a number of iterations until the deposition levels stabilise.

The basic model can be modified to remove the deposits from the atmosphere and to resemble more closely the actual system.

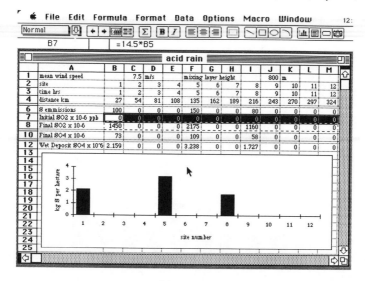

Figure 15.2 *Cells B8 to M8 have been highlighted and copied from the edit menu and are now designated by the broken border. Cells B7 to M7 have been highlighted to receive the data*

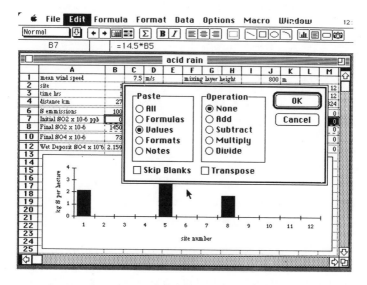

Figure 15.3 *'Paste Special' within the edit menu has been selected so that only the values of the final concentrations are pasted into the initial concentration cells to begin the next iteration and not the actual formulae*

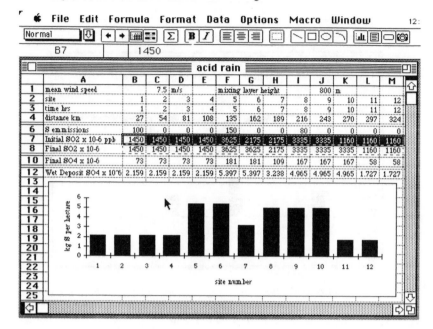

Figure 15.4 *A number of iterations have now been completed and the wet deposition levels are equilibrating in a downwind direction*

REFERENCES

Alcamo, J., Amann, M., Hettelingh, J-P., Holmberg, M., Hardijk, L., Kamari, J., Kauppi, L., Kornai, G. & Makela, A., 1987. Acidification in Europe: A simulation model for evaluating control strategies. *Ambio*, **16**, 232–245.

Battarbee, R.W., Flower, R.J., Stevenson, A. & Rippey, B., 1985. Lake acidification in Galloway: A palaeoecological test of competing hypotheses. *Nature*, **314**, 350–352.

CEC, 1988. Commission of the European Communities. Council Directive on the limitations of emissions of certain pollutants into the air from large combustion plants, 88/609/EC. *Official Journal of the European Communities No.L 336*, **31**, 1–13.

Chang, T.S, Brost, R.A., Isaksen, I., Madhonick, S., Middleton, P., Stockwell, W.R. & Walek, C., 1987. A 3-D Eulerian acid deposition model—physical concepts and formulation. *Journal of Geophysical Research*, **92**, 14681–14700.

Eliassen, A. & Saltbones, J., 1983. Modelling of long-range transport of sulphur over Europe: A two year model run and some model experiments. *Atmospheric Environment*, **17**, 1457–1473.

Fisher, B.E., 1978. The calculation of long-term sulphur deposition in Europe. *Atmospheric Environment*, **12**, 489–501.

Hough, A.M. & Eggleton, A.E., 1986. Acid deposition modelling: A discussion of the processes involved and the options for future development. *Science Programme Oxford*, **70**, 353–379.

Metcalfe, S.E. & Derwent, R.G., 1990. Llyn Brianne—Acid deposition modelling. In: R.W. Edwards, A.S. Gee and J.H. Stoner (eds.) *Acid Waters in Wales*. Kluwer, Dordrecht, pp.299–309.

Metcalfe, S.E., Atkins, D. & Derwent, R.G., 1989. Acid deposition modelling and the interpretation of the United kingdom secondary precipitation network data. *Atmospheric Environment*, **23**, 2033–2052.

Nordic Council of Ministers, 1986. *Europe's Air, Europe's Environment*. Norstedts Tryckeri, Stockholm (English translation).

RGAR, 1990. *Acid Deposition in the United Kingdom 1986–1988*. DoE/Warren Spring Laboratory.

16
Hydrological Response of Lake Basins

INTRODUCTION

It has been recognised for more than 250 years that lake levels fluctuate in response to climatic change (Halley, 1715; cited in Street-Perrott & Harrison, 1985). This sensitivity has assumed renewed significance in the light of current concerns over global climatic change induced by an enhanced greenhouse effect (Jager & Ferguson, 1991). Lakes perform an important role in the functioning of the hydrological system, mediating the response of catchments to precipitation inputs and often providing a major resource base in terms of fishing, navigation and as a source of water. This role is particularly significant in arid and semi-arid areas where lakes provide the only perennial source of water.

The association between climatic change and lake level change over a range of time scales from years to millenia has been extensively reviewed elsewhere. Over the period of the Pleistocene–Holocene (last glacial to present interglacial period), evidence of lake level change provided by geomorphological (e.g. old lake shorelines) and sedimentological (e.g. facies changes, fossil microfauna and microflora) data, has played an important part in reconstructing climatic change. This has been particularly true in areas of the dry tropics and sub-tropics which sustained extensive lakes at varying times over the last glacial cycle. The timing of the regional high stands was very variable, depending upon shifts in the atmospheric circulation (Street-Perrott & Harrison, 1984). Lakes in the East African Rift Valley, for example, were low at the time of the glacial maximum and late glacial (ca. 18 000–12 000 yr BP), but high in the early to mid Holocene under the influence of an enhanced summer monsoon circulation round the Indian Ocean (Street-Perrott & Roberts, 1983). By contrast, the lake basins of the south-west USA were full at the last glacial maximum due to a southward displacement of the mid-latitude Westerlies by the huge mass of the Laurentide ice sheet (Smith & Street-Perrott, 1983; Harrison & Metcalfe, 1985). The spatial and temporal patterns of lake level change have provided important checks for atmospheric global circulation models attempting to reconstruct past conditions (COHMAP member, 1988). Over the historical period, the details of this link become clearer on an annual or decadal basis. Sirculon (1987) has examined the nineteenth and twentieth century records for rivers and lakes in West Africa in relation to climate. In this area, the tropical rivers bring vital water resources into the Sahelian zone and Lake Chad. The effects of the Sahelian drought from the mid 1970s to mid 1980s are particularly clear. The water balance of Lake Chad at the present and in the past has been studied in more detail by Kutzbach (1980) and Tetzlaff and Adams (1983).

Street (1980) made an assessment of the relative sensitivity of different types of lakes to changes in climate. Lakes may be divided into two basic categories:

1 closed lakes having no overflow or losses through groundwater seepage, and
2 open lakes which have either a surface outlet or lose water through seepage although topographically closed.

Of these two categories, closed basin lakes are by far the more sensitive to climatic change, undergoing large fluctuations in level in response to changes in the balance of $P–Et$ (precipitation–evapotranspiration). Where groundwater effects are negligible,

Street recognises three types of lakes: flow-dominated reservoirs, atmosphere-controlled lakes and amplifier lakes. Amplifier lakes, with a water balance dominated by surface runoff from the catchment and by evaporation, are the most responsive to changes in climate, and undergo major changes in water level. Atmosphere-controlled lakes also respond directly to changes in water balance (P–Et), although less sensitively. It is, therefore, lakes which fall into one of these two latter categories which are most suitable for exploring the effects of climatic change on lake water balance. Another important factor affecting sensitivity is basin relief, with a small basin area:depth ratio being the best (Street-Perrott & Harrison, 1985).

According to the above authors, the mean annual water balance of a lake can be expressed as:

$$V = A_L(P_L - E_L) + (R - D) + (G_I - G_O)$$

where

V = change in volume,
A_L = lake area,
P_L = precipitation over the lake,
E_L = evaporation from the lake,
R = runoff from the catchment,
D = surface discharge from the lake,
G_I = groundwater flow in to the lake, and
G_O = groundwater flow out of the lake.

For a closed basin lake, $D = 0$, and if you assume that G_I and G_O are negligible and that equilibrium conditions prevail, then the area of a closed lake can be seen to be dependent upon precipitation and evaporation over its surface and catchment.

Through manipulating the water balance equation for equilibrium conditions it is possible, given the availability of different types of information, to calculate either: (a) precipitation, if the palaeolake area is known (e.g. from shorelines) and if you estimate E_L and either R or E_B; or (b) lake area from estimates of other variables. The applications of this are clear with (a) being useful for interpreting geological/historical evidence and (b) allowing for prediction of change under future climates. An example of the former is Street's (1977) paper which estimates Late Quaternary precipitation in southern Ethiopia, and of the latter, Sanderson and Wong's (1987) study of the Great Lakes. It is interesting to note that the level of the Great Lakes is responsive to climatic change even though they are hydrologically open. These authors point out that whilst people may be able to adapt to the 'usual' range of lake level change (2 m in the case of the Great Lakes), future global climate change might lead to fluctuations larger than this, posing additional problems. Looking to the future, Kotwicki and Clark (1991) have suggested that lakes, primarily those in arid and semi-arid areas could be sensitive indicators of emerging climatic change.

Gleick (1989) has proposed a three-step methodology to evaluate the impact of climatic changes on the water cycle:

(1) Develop quantitative scenarios of changes in major climatic variables such as temperature, precipitation and radiation (e.g. through the use of GCMs, see Chapter 2).
(2) Using the climatic parameters developed in (1), simulate the hydrological cycle of the basin being studied. Potential and actual evapotranspiration, soil moisture, runoff, etc. can be calculated using hydrological models.
(3) Assess the implications of the hydrological changes identified in (2) for water resources, e.g. water supply, irrigation.

This methodology can provide a framework for the use of simple lake water balance models for the purposes of prediction.

MODELLING WATER BALANCE FOR LAKES

The two basic approaches to modelling water balance for a lake are to use a simple water balance model, or a combined water–energy balance model (Kutzbach, 1980; Street-Perrott & Harrison, 1985). Simple water balance models are very sensitive to the value chosen for evaporation over the lake (E_L). Such estimates are usually based on palaeotemperature reconstructions from tree lines and other vegetation evidence, periglacial features, etc. Even in the present day, measuring evaporation is difficult. Kotwicki and Clark (1991) report evaporation values for Lake Eyre (Australia) ranging between 1800 mm and 3600 mm depending upon the method used. Estimating runoff, or a runoff coefficient, is also difficult, usually being calculated from current measurements of rainfall, runoff and/or evapotranspiration. Published runoff figures can be used. Such calculations usually make little allowance for changes in vegetation, soils and groundwater (Street, 1980).

Combined water–energy balance models allow palaeoevaporation to be calculated in a more sophisticated way, allowing for fluxes of latent and sensible heat and net radiation flux. With this approach, factors such as cloud cover, cloud type and windspeed also need to be considered.

Modelling lake levels is based upon a relatively simple mass continuity equation for water (e.g. Street-Perrott & Harrison, 1985) which is solved for an equilibrium lake area as described here. The mean annual water balance for a lake is given by:

$$V = A_L(P_L - E_L) + (R - D) + (G_I - G_O) \tag{16.1}$$

where:

V is the net change in volume of the lake (m^3)

A_L is the lake area (m^2),

P_L is annual precipitation into the lake ($m\ m^{-2}$),

E_L is the annual evaporation from the lake ($m\ m^{-2}$),

R is the runoff from the catchment (m^3),

D is the discharge from the lake (m^3),

G_I is the groundwater input into the lake (m^3), and

G_O is the groundwater output from the lake (m^3).

For closed lakes, D is zero and under equilibrium conditions (i.e. $V = 0$) and assuming that groundwater transfers are negligible, equation (16.1) reduces to give the runoff input as equal to the difference between evaporation and precipitation for the whole of the surface area of the lake:

$$R = A_L(E_L - P_L) \tag{16.2}$$

If it is further assumed that runoff from the drainage basin can be represented by the difference between drainage basin precipitation and evaporation (P_B and E_B respectively), for a basin area of A_B:

$$R = A_B(P_B - E_B) \qquad (16.3)$$

then, equations (16.2) and (16.3) can be combined to determine the equilibrium area of the lake:

$$A_L = A_B \frac{(P_B - E_B)}{(E_L - P_L)} \qquad (16.4)$$

Finally, the problem can be further simplified by assuming that precipitation is the same for both lake and catchment basin ($P_B = P_L$) so that equation (16.4) becomes:

$$A_L = A_B \frac{(P_B - E_B)}{(E_L - P_B)} \qquad (16.5)$$

Equation (16.5) states that precipitation must be greater than evaporation within the basin as whole so that water is being input into the lake for a stable body of water to form and that evaporation must be greater than precipitation so that excess water is removed. Equation (16.5) forms the basis of the model described in the following section.

MODEL IMPLEMENTATION

The screen display for the lake model is shown in Figure 16.1. The control panel occupies B2 to D2 and a small text window has been inserted, using the text button, to contain the warning that cell C3 (that is the ratio of basin evaporation to basin precipitation) must be less than one in order for a viable model to result. In the real world, as lakes are in the process of desiccation, this ratio could be more than

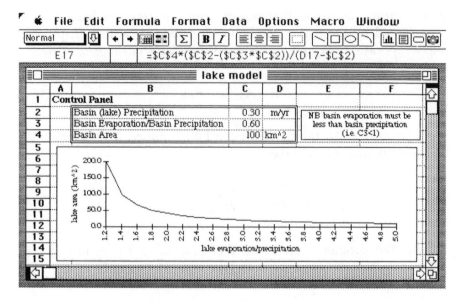

Figure 16.1 *Reduction in lake area with increase in the evaporation/precipitation ratio*

one. The actual engine of the model is contained in cells C17, D17 and E17 and downwards, which calculate the value of evaporation from the lake (i.e. E_L is in D17, etc.) and then the equilibrium lake area in column E. The formula is shown in the active cell reference and formula bar at the top of the screen. The resulting lake areas for a range of evaporation to precipitation ratios are plotted in the graph.

EXAMPLES

Any number of 'what if' scenarios can be explored with this simple model. Figure 16.2, for example, demonstrates the effect of increasing the evaporation to precipitation ratio, Figure 16.3, the effect of increasing the basin area; and Figure 16.4, the effect of decreasing the basin precipitation. It becomes likely that the equilibrium solution for lake area operates in some kind of dynamic equilibrium and the model could readily be modified to demonstrate this point.

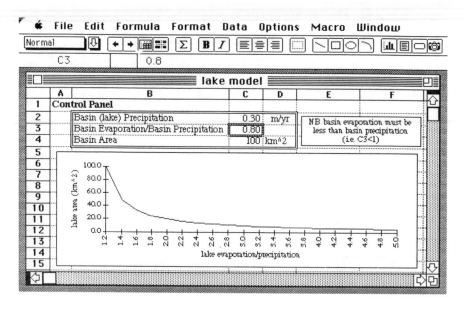

Figure 16.2 *Effect of increasing the input ratio*

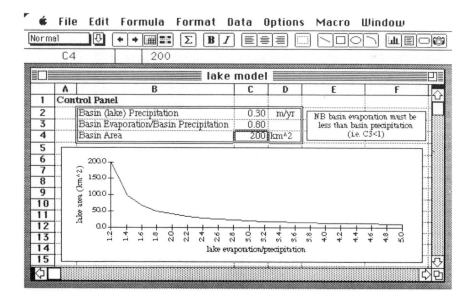

Figure 16.3 *Effect of increasing the basin area*

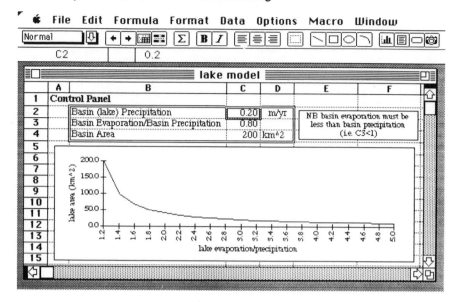

Figure 16.4   *Effect of decreasing the precipitation input*

REFERENCES

COHMAP members, 1988. Climatic changes of the last 18,000 years: observations and model simulations. *Science*, **241**, 1043–1052.

Gleick, P.H., 1989. Climate change, hydrology and water resources. *Reviews of Geophysics*, **27**, 329–344.

Harrison, S.P & Metcalfe, S.E., 1985. Spatial variations in lake levels since the last glacial maximum in the Americas north of the equator. *Zeitschrift fur Gletscherkunde und Glazialgeologie*, **21**, 1–15.

Jager, J. & Ferguson, H.L., 1991 *Climate Change: Science, Impacts and Policy*. Cambridge University Press, Cambridge.

Kotwicki, V. & Clark, R., 1991. Aspects of the water balance of three Australian terminal lakes. In: *Hydrology of Natural and Manmade Lakes*, International Association of Hydrological Sciences Publication 206 pp.3–12.

Kutzbach, J.E., 1980. Estimates of past climates at Paleolake Chad, north Africa based on a hydrological and energy balance model. *Quaternary Research*, **14**, 210–223.

Kutzbach, J.E., 1983. Monsoon rains of the Late Pleistocene and Early Holocene: patterns, intensity and possible causes of change. In: A. Street-

Perrott, M. Beran & R. Ratcliffe (eds) *Variations in the Global Water Budget.* D. Reidel, Dordrecht pp. 371–389.

Sanderson, M. & Wong, L., 1987. Climatic change and Great Lakes water levels. In: *The Influence of Climatic Change and Climatic variability on the Hydrologic Regime and Water Resources*, International Association of Hydrological Sciences Publication 168, pp. 477–487.

Sirculon, J., 1987. Variation des débits des cours d'eau et des niveaux des lacs en Afrique de l'ouest depuis le début du 20 ième siècle. In: *The Influence of Climatic Change and Climatic Variability on the Hydrologic Regime and Water Resources.* International Association of Hydrological Sciences Publication 168, pp. 13–25.

Smith, G.I. & Street-Perrott, F.A., 1983. Pluvial lakes of the western United States. In: S.C. Porter (ed.) *Late Quaternary Environments of the United States. Vol. 1, The Late Pleistocene.* University of Minnesota Press, Minneapolis, pp. 190–212.

Street, F.A., 1977. Late Quaternary precipitation estimates for the Ziway-Shala basin, southern Ethiopia. *Palaeoecology of Africa*, **10/11**, 135–143.

Street, F.A., 1980. The relative importance of climate and local hydrogeological factors in influencing lake-level fluctuations. *Palaeoecology of Africa*, **12**, 137–158.

Street-Perrott, F.A. & Harrison, S.P., 1984. Temporal variations in lake levels since 30,0000 yr BP—an index of the global hydrological cycle. In: *Climate Processes and Climate Sensitivity*, vol.5. Geophysical Monograph 29, Maurice Ewing, pp. 118–129.

Street-Perrott, F.A. & Harrison, S.P., 1985. Lake levels and climate reconstruction. In: A.D. Hecht (ed.) *Paleoclimate Analysis and Modeling.* J. Wiley & Sons, New York, pp. 291–340.

Street-Perrott, F.A. & Roberts, N., 1983. Fluctuations in closed-basin lakes as an indicator of past atmospheric circulation patterns. In: A. Street-Perrott, M. Beran & R. Ratcliffe (eds) *Variations in the Global Water Budget.* D. Reidel, Dordrecht, pp. 331–345.

Tetzlaff, G. & Adams, L.J., 1983. Present-day and early-Holocene evaporation of Lake Chad. In: A. Street-Perrott, M. Beran & R. Ratcliffe (eds) *Variations in the Global Water Budget.* D. Reidel, Dordrecht pp. 347–360.

Part IV
APPENDICES

Appendix I
The Model.Blank File

INTRODUCTION

Excel has been used throughout this book and, in particular, the screen shots have been taken from the Apple Macintosh version of the application. Although the software and models described here are available in both DOS and Macintosh formats, readers may wish to construct their own models without reference to the disk. The Model.blank file which is mentioned in Chapter 4 can form a basis for this work and is shown overleaf (Figure AI.1). It can be built by opening the spreadsheet and then arranging the row heights and column widths so that all of columns A to C and rows 1 to 19 appear on the screen. In addition, integer format is used throughout the screen, so highlight column C and select integers from the Format Number command.

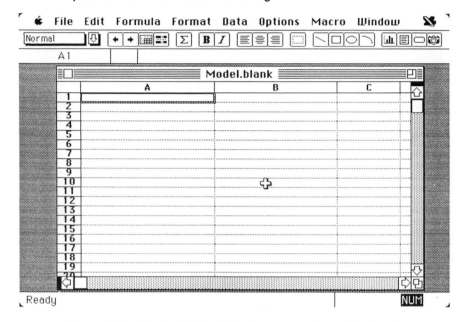

Figure AI.1   *The Model.blank file when opened on a Macintosh*

Appendix II
Excel Functions

INTRODUCTION

Excel has been progressively introduced through the earlier chapters in this book and a number of functions have been explained and utilised. There are, however, a large number of other functions which can be used within formulae in Excel and these represent the real programming power of the application. The following lists all such functions in an abbreviated, ready reference form and the user should refer to the reference manuals for further details. The functions are divided into database (which has not been covered in this book), date and time, financial functions, information, logical, lookup, mathematical, matrix, staistical, text and trigonometric functions. The best advice is to try them out and see if they produce the desired result.

DATABASE FUNCTIONS

DAVERAGE(*database,field,criteria*)
  Returns the average of selected database entries
DCOUNT(*database,field,criteria*)
  Counts the cells containing numbers from a specified database and criteria
DCOUNTA(*database,field,criteria*)
  Counts nonblank cells from a specified database and criteria

DGET(*database,field,criteria*)

Extracts from a database a single record that matches the specified criteria

DMAX(*database,field,criteria*)

Returns the maximum value from selected database entries

DMIN(*database,field,criteria*)

Returns the minimum value from selected database entries

DPRODUCT(*database,field,criteria*)

Multiplies the values in a particular field of records that match the criteria in a database.

DSTDEV(*database,field,criteria*)

Estimates the standard deviation based on a sample of selected database entries.

DSTDEVP(*database,field,criteria*)

Calculates the standard deviation based on the entire population of selected database entries.

DSUM(*database,field,criteria*)

Adds numbers in a database

DVAR(*database,field,criteria*)

Estimates variance based on a sample from selected database entries

DVARP(*database,field,criteria*)

Calculates variance based on the entire population of selected database entries

DATE AND TIME FUNCTIONS

DATE(*year,month,day*)

Returns the serial number of a specified date

DATEVALUE(*date_text*)

Converts a date in the form of text to a serial number

DAY(*serial_number*)

Converts a serial number to a day

DAYS360(*start_date,end_date*)

Calculates the number of days between two dates based upon a 360-day year

HOUR(*serial_number*)

Converts a serial number to an hour

MINUTE(*serial_number*)

Converts a serial number to an minute

MONTH(*serial_number*)

Converts a serial number to an month

NOW()

Returns the serial number of the current date and time

SECOND(*serial_number*)

Converts a serial number to a second

TIME(*hour_minute_second*)

Returns the serial number of a particular time

TIMEVALUE(*time_text*)

Converts a time in the form of text to a serial number

TODAY()

Returns the serial number of today's date

WEEKDAY(*serial_number*)

Converts a serial number to a day of the week

YEAR(*serial_number*)

Converts a serial number to a year

FINANCIAL FUNCTIONS

DDB(*cost,salvage,life,period,factor*)

Returns the depreciation of an asset for a specified period using the double-declining balance method

FV(*rate,nper,pmt,pv,type*)

Returns the future value of an investment

IPMT(*rate,per,nper,pv,fv,type*)

Returns the interest payment for an investment for a given period

IRR(*values,guess*)

Returns the internal rate of return for an investment, without financing costs or reinvestment gains

MIRR(*values,finance_rate,reinvest_rate*)

Returns the internal rate of return where positive and negative cash flows are financed at different rates

NPER(*rate,pmt,pv,fv,type*)

Returns the number of payments for an investment

NPV(*rate,value1,value2,...*)

Returns the net present value of an investment based on cash flows that do not have to be constant

PMT(*rate,nper,pv,fv,type*)

Returns the periodic total payment for an investment

PPMT(*rate,per,nper,pv,fv,type*)

Returns the payment on the principle for an investment for a given period

PV(*rate,nper,pmt,fv,type*)

Returns the present value of an investment

RATE(*nper,pmt,pv,fv,type,guess*)

Returns the interest rate per period of an investment

SLN(*cost,salvage,life*)

Returns the straight-line depreciation of an asset for one period

SYD(*cost,salvage,life,per*)

Returns the sum-of-years' digits depreciation of an asset for a specified period

VDB(*cost,salvage,life,start_period,end_period,factor,no_switch*)

Returns the depreciation of an asset for a specified or partial period using a declining balance method

INFORMATION FUNCTIONS

ADDRESS(*row_num,column_num,abs_num,a1,sheet_text*)

Returns a reference as text to a single cell in a worksheet

AREAS(*reference*)

Returns the number of areas in a reference

CELL(*info_type,reference*)

Returns information about the formatting, location, or contents of a cell

COLUMN(*reference*)

Returns the column number of a reference

COLUMNS(*array*)

Returns the number of columns in a reference

INDIRECT(*ref_text,a1*)

Returns a reference indicated by a text value

INFO(*type_num*)

Returns information about the current operating environment

ISBLANK(*value*)

Returns 'TRUE' if the value is blank

ISERR(*value*)

Returns 'TRUE' if the value is any error value except #N/A

ISERROR(*value*)

Returns 'TRUE' if the value is any error value

ISLOGICAL(*value*)

Returns 'TRUE' if the value is a logical value

ISNA(*value*)

Returns 'TRUE' if the value is the error value #N/A

ISNONTEXT(*value*)

Returns 'TRUE' if the value is not text

ISNUMBER(*value*)

Returns 'TRUE' if the value is a number

ISREF(*value*)

Returns 'TRUE' if the value is a reference

ISTEXT(*value*)

Returns 'TRUE' if the value is text

N(*value*)

Returns the value converted to a number

NA()

Returns the error value #N/A

OFFSET(*reference,rows,cols,height,width*)

Returns a reference offset from a given reference

ROW(*reference*)

Returns the row number of a reference

ROWS(*array*)

Returns the number of rows in a reference

T(*value*)

Converts its arguments to text

TYPE(*value*)

Returns a number indicating the data type of a value

LOGICAL FUNCTIONS

AND(*logical1,logical2,...*)

Returns 'TRUE' if all its arguments are TRUE

FALSE()

Returns the logical value 'FALSE'

IF(*logical_test,value_if_true,value_if_false*)

Specifies a logical test to perform

NOT(*logical*)

Reverses the logic of its argument

OR(*logical1,logical2,...*)

Returns 'TRUE' if any argument is TRUE

TRUE()

Returns the logical value 'TRUE'

LOOKUP FUNCTIONS

CHOOSE(*index_num,value1,value2,...*)

Chooses a value from a list of values

HLOOKUP(*lookup_value,table_array,row_index_num*)

Looks in the first row of an array and moves down the column to return
the value of a cell

INDEX(*reference,row_num,column_num,area_num*)
INDEX(*array,row_num,column_num*)
  Uses an index to choose a value from a reference or array
LOOKUP(*lookup_value,lookup_vector,result_vector*)
LOOKUP(*lookup_value,array*)
  Looks up values in an array or reference
MATCH(*lookup_value,lookup_array,match_type*)
  Looks up values in a reference or array
VLOOKUP(*lookup_value,table_array,col_index_num*)
  Looks in the first column of an array and moves across the row to return
  the value of a cell

MATHEMATICAL FUNCTIONS

ABS(*number*)
  Returns the absolute value of a number
EXP(*number*)
  Returns e raised to a given number
FACT(*number*)
  Returns the factorial of a number
INT(*number*)
  Rounds a number down to the nearest integer
LN(*number*)
  Returns the natural logarithm of a number
LOG(*number,base*)
  Returns the logarithm of a number to a specified base
LOG10(*number*)
  Returns the base-10 logarithm of a number
MOD(*number,divisor*)
  Gives the remainder from division
PI()
  Returns the value π
PRODUCT(*number1,number2,...*)
  Multiplies its arguments
RAND()
  Returns a random number between 0 and 1
ROUND(*number,num_digits*)
  Rounds a number to a specified number of digits
SIGN(*number*)
  Returns the sign of a number

SQRT(*number*)
   Returns the square root of a number
SUM(*number1,number2. . .*)
   Adds its arguments
TRUNC(*number,num_digits*)
   Truncates a number to an integer

MATRIX FUNCTIONS

MDETERM(*array*)
   Returns the matrix determinant of an array
MINVERSE(*array*)
   Returns the matrix inverse of an array
MMULT(*array1,array2*)
   Returns the matrix product of an array
SUMPRODUCT(*array1,array2*)
   Multiplies corresponding components in the given arrays and returns the
   sum of those products
TRANSPOSE(*array*)
   Returns the transpose of an array

STATISTICAL FUNCTIONS

AVERAGE(*number1,number2,. . .*)
   Returns the average of its arguments
COUNT(*value1,value2,. . .*)
   Counts how many numbers are in the list of arguments
COUNTA(*value1,value2,. . .*)
   Counts how many values are in the list of arguments
GROWTH(*known_y's,known_x's,new_x's,const*)
   Returns values along an exponential trend $y = b * m^x$
LINEST(*known_y's,known_x's,const,stats*)
   Returns the parameters of a linear trend $y = m_1x_1 + m_2x_2 + \ldots + b$ or
   $y = m * x + b$
LOGEST(*known_y's,known_x's,const,stats*)
   Returns the parameters of an exponential trend $y = (b * (m_1 {}^\wedge x_1) * (m_2 {}^\wedge x_2) *$
   $\ldots)$ or $y = b * m^\wedge x$
MAX(*number1,number2,. . .*)
   Returns the maximum value in a list of arguments

MEDIAN(*number1,number2,. . .*)

   Returns the median of its arguments

MIN(*number1,number2,. . .*)

   Returns the minimum value in a list of arguments

STDEV(*number1,number2,. . .*)

   Estimates standard deviation based on a sample

STDEVP(*number1,number2,. . .*)

   Calculates standard deviation based on the entire population

TREND(*known_y's,known_x's,new_x's,const*)

   Returns values along a linear trend $y = m * x + b$

VAR(*number1,number2,. . .*)

   Estimates variance based on a sample

VARP(*number1,number2,. . .*)

   Calculates variance based on the entire population

TEXT FUNCTIONS

CHAR(*number*)

   Returns the character specified by the code number

CLEAN(*text*)

   Removes control characters from text

CODE(*text*)

   Returns the code number of the first character in text

DOLLAR(*number,decimals*)

   Formats a number and converts it to text

EXACT(*text1,text2*)

   Checks to see if two text values are identical

FIND(*find_text,within_text,start_at_num*)

   Finds one text value within another (case sensitive)

FIXED(*number,decimals*)

   Formats a number as text with a fixed number of decimals

LEFT(*text,num_chars*)

   Extracts the leftmost characters from a text value

LEN(*text*)

   Returns the length of a text string

LOWER(*text*)

   Converts text to lower case

MID(*text,start_num,num_chars*)

   Extracts a number of characters from text

PROPER(*text*)

   Converts text to initial capitals

REPLACE(*old_text,start_num,num_chars,new_text*)

Replaces characters within text

REPT(*text,number_times*)

Repeats text a given number of times

RIGHT(*text,num_chars*)

Extracts the rightmost characters from a text value

SEARCH(*find_text,within_text,start_num*)

Finds one text value within another (not case sensitive)

SUBSTITUTE(*text,old_text,new_text,instance_num*)

Replaces characters within text

TEXT(*value,format_text*)

Formats a number and converts it to text

TRIM(*text*)

Removes spaces from text

UPPER(*text*)

Capitalizes every letter in a text value

VALUE(*text*)

Converts a text argument to a number

TRIGONOMETRIC FUNCTIONS

ACOS(*number*)

Returns the arccosine of a number

ACOSH(*number*)

Returns the inverse hyperbolic cosine of a number

ASIN(*number*)

Returns the arcsine of a number

ASINH(*number*)

Returns the inverse hyperbolic sine of a number

ATAN(*number*)

Returns the arctangent of a number

ATAN2(*x_num,y_num*)

Returns the arctangent from *x*- and *y*-coordinates

ATANH(*number*)

Returns the inverse hyperbolic tangent of a number

COS(*number*)

Returns the cosine of a number

COSH(*number*)

Returns the hyperbolic cosine of a number

SIN(*number*)

Returns the sine of a number

SINH*(number)*

  Returns the hyperbolic sine of a number

TAN*(number)*

  Returns the tangent of a number

TANH*(number)*

  Returns the hyperbolic tangent of a number

Appendix III
Works, Lotus and Other Spreadsheets

INTRODUCTION

Although this books has been written around the Microsoft Excel applications running on Apple Macintosh machines, there are a number of other, well-known spreadsheets which could equally well be utilised. Since all Macintosh application and DOS applications running under a mouse-controlled DOS.SHELL or in Windows look and feel very similar, the differences are becoming smaller and less important. Indeed, most of the more well-known spreadsheets are, in many instances, file compatible so that a model developed in one will often load and run in another. As an example, the diagram overleaf (Figure AIII.1) shows an opening screen of the integrated Microsoft Works package which includes a spreadsheet along with wordprocessor and database. Clearly, the similarities with Excel are apparent and a few minutes of experimentation will show that most of the exercises described in this book can be accomplished in most other spreadsheets.

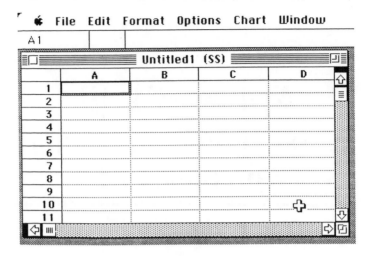

Figure AIII.1

Index

NOW AVAILABLE

The programs described in this book are available on disk for your IBM PC compatible and Macintosh computers, using Excel versions 2, 3 or 4.

The disk contains the six models described in chapters 11 to 16 of this book.

Order the Program Disk today priced £9.95 including VAT/$15.95 from your computer store, bookseller, or by using the order form below.

If you have any queries about the compatibility of your
hardware configuration please contact:

Helen Ramsey
John Wiley & Sons Ltd
Baffins Lane
Chichester
West Sussex
PO19 1UD
England

Customer Service Department
John Wiley & Sons Ltd
Shripney Road
Bognor Regis
West Sussex
PO22 9SA
England